Mass Flow Measurement

Mass Flow Measurement describes how Coriolis flowmeters, thermal flowmeters, mass flow controllers, and multivariable flowmeters function. It explains how and why these flowmeters were developed, illustrates how they work and the applications and ways in which they are used, and explores what to expect in terms of future development.

Covering flowmeters that specifically measure mass flow, the book discusses why flow measurement is important for measuring gas flow and any fluid whose quantity is affected by temperature and pressure. Coriolis mass flow controllers measure mass directly, but many also contain a valve for control purposes, as they are widely used in semiconductor and industrial environments. It also introduces multivariable flowmeters, which measure volumetric flow and then compensate for temperature and pressure.

This book is designed for professionals involved with flowmeters and instrumentation, including product and marketing managers, strategic planners, application engineers, and distributors.

Jesse Yoder, President of Flow Research, Inc., USA, has 36 years of experience writing about and analyzing process control and instrumentation markets, beginning as President and Founder of Idea Network, USA. He received a B.A. degree in philosophy at the University of Maryland, USA, in 1973 and a PhD in philosophy from the University of Massachusetts, USA in 1984. Dr. Yoder was an adjunct philosophy professor at the University of Massachusetts, Lowell, USA and Lafayette College, USA for 10 years. He founded Flow Research, Inc. in 1998, where he still serves as the President. Flow Research is a market research company that publishes market studies on flow and instrumentation topics. Dr. Yoder has received two patents for new dual-sensor flowmeter designs. Several prototypes of these designs have been built and tested at CEESI. Dr. Yoder has prepared over 300 market studies, published over 300 articles on flow and instrumentation in industry journals, and has written several books, including *New-Technology Flowmeters* (CRC Press, 2023) and *Conventional Flowmeters* (CRC Press, 2023).

Mass Flow Measurement

Jesse Yoder

CRC Press
Taylor & Francis Group
Boca Raton London New York

CRC Press is an imprint of the
Taylor & Francis Group, an **informa** business

Designed cover image: Jesse Yoder

First edition published 2026
by CRC Press
2385 NW Executive Center Drive, Suite 320, Boca Raton FL 33431

and by CRC Press
4 Park Square, Milton Park, Abingdon, Oxon, OX14 4RN

CRC Press is an imprint of Taylor & Francis Group, LLC

ISBN: 978-1-032-60160-1 (hbk)
ISBN: 978-1-032-60163-2 (pbk)
ISBN: 978-1-003-45785-5 (ebk)

DOI: 10.1201/9781003457855

Typeset in Times
by KnowledgeWorks Global Ltd.

Access the Support Material: Routledge.com/9781032601601

To Belinda, who made it all happen.

Contents

Preface

Edited by Belinda Busteed Burum.

There are many concepts in flow that I have struggled for many years to get my head around. One is mass flow, another is "the Coriolis principle," and still another is how it is possible to measure the area of a pipe using π number of squares. I have puzzled over the Coriolis principle and mass flow for over 20 years. I have puzzled over the formula for measuring the area of a circle for over 40 years. Mass is a slightly more recent puzzle.

Although the main aim of this book is to explore the concept of mass flow measurement in depth, I have tried to address all three issues in this book, as they are all related. I find that the best way for me to solve a problem is to write about it. That doesn't mean I solve the problem on my first attempt. In the case of mass flow, I believe I have made significant progress in this book towards understanding what it is at a fairly deep level, though some puzzles still remain.

DOES A CORIOLIS METER REALLY USE THE CORIOLIS PRINCIPLE?

The Coriolis principle occupies a significant portion of this book. I've published many articles on the Coriolis principle and on Coriolis flowmeters. I've done many market studies on this subject, including custom projects. Yet I still wondered if and why the principle used in Coriolis flowmeters should be called the Coriolis principle. To help answer this question, I went back to the 1835 manuscript by Gaspard Gustave de Coriolis that started the flowmeter world's discussion about Coriolis force and Coriolis effect. Because this manuscript was written in French and had not previously been translated, I worked with Belinda Burum to arrive at the first English translation of the manuscript. I also found that Coriolis's 1835 article doesn't mention waterwheels, although it is about the analysis of motion in a rotating frame of reference.

Many modern summaries of Gaspard-Gustave de Coriolis's work state that his 1835 paper on rotating systems was inspired by waterwheels, or that he was talking about waterwheels in this 1835 article. An examination of the text, after its translation, reveals that he doesn't refer to waterwheels at all in this article. The idea that Coriolis was writing about waterwheels is a later analogy designed to make the article more understandable (see Appendix).

I had lengthy discussions with top Coriolis experts at Emerson, KROHNE, Endress+Hauser, and other Coriolis companies. This included multiple in-person visits to the home offices of these companies, both in the United States and in Europe. I am grateful to these companies for their hospitality and for making available their best people to discuss the Coriolis principle with me. My quest to understand this principle has been an iterative one, making small steps and analyzing the topic from many points of view.

Today, the term "Coriolis effect" is often associated with weather systems, although it was not until 1901 that it was used in a meteorological context. As to how the name Coriolis came to be applied to Coriolis meters, it appears to be an accident of history. In 1958, Roby White was granted a patent that mentioned the "Coriolis force." In 1960, Yao Tzu Li was granted a patent called "Mass Flowmeter." Both these patents proposed rotating the fluid. Li's patent was an improvement on White's 1958 design. In 1965, Anatole Sipin concluded that rotating the fluid was impractical, and received a patent for a flowmeter that oscillated the fluid instead of rotating it. Despite this, he referred to the "Coriolis force" in his patent. Later, in his 1978 patent, Jim Smith kept the idea of oscillation and called his meters Coriolis meters.

While many people with technical minds superior to my own continue to refer to the Coriolis force as an explanation for the operation of Coriolis flowmeters, I have come to the conclusion that that this operation can also be understood in terms of the effect of inertial forces, rather than the Coriolis force. Oscillation is not rotation, and it is rotation, not oscillation, that Coriolis talks about in his 1835 article. The deflection of the flowtube in a Coriolis meter can be described as the result of the force exerted by the inertial mass of the fluid. The phase difference (time delay) between the sine waves produced by the inlet and outlet sensors is directly proportional to the mass flowrate of the fluid. A larger time delay indicates a higher mass flowrate. This statement is true whether we attribute the time delay to the action of inertial forces or to Coriolis forces.

Perhaps the best way to look at the question about inertial vs. Coriolis force is as two different frameworks. The key value used to compute mass flow is the phase difference (time delay) between the sine waves produced by the inlet and outlet sensors. This phase difference (time delay) is produced as a result of the deflection ("twisting") of the flowtubes. The question is whether this deflection of the flowtubes is caused by Coriolis forces or by the inertia of the fluid. If this deflection is caused by Coriolis forces, then they should be seen in terms of the framework described by Gaspard Gustave de Coriolis in his relevant articles (primarily the 1835 article). However, we have already seen that he was talking about a rotating frame of reference, not about oscillation. If the deflection is caused by inertial forces, there is no need to appeal to the Coriolis force or Coriolis effect to explain the deflection. This book discusses this topic in more detail.

WHAT IS MASS, REALLY?

I find mass to be both an intuitive and an elusive concept. Mass seems obvious, and yet it is not. In writing this book, I really wanted to understand what mass is, and it led me down the proverbial rabbit hole to the depths of subatomic particles.

I began with the idea that mass is the amount of matter in an object. I then looked more carefully at what an object is and drew on Bertrand Russell who says essentially, "the world is not as it seems." We experience objects like rocks, houses, and desks as being stable, perceivable by more than one person, and as existing through time. Scientists, on the other hand, view these same objects as collections of atoms and molecules that only appear solid to us because they are moving so fast. To follow this logic, then, I traced the concept of atoms through to the atomic table of elements. But even atoms themselves are not simple and indivisible, as Democritus thought. Instead, atoms are composed of protons, neutrons, and electrons. And the bulk of the mass of an atom comes from the protons and neutrons, with very little coming from the electrons. However, even protons are not indivisible. Instead, they are composed of quarks, gluons, and other subatomic particles moving at incredibly high speeds. It is here that we find the secret of mass: subatomic particles moving at near-light speed.

Happily, I'm not alone in my desire to understand the mystery of mass or matter. The search goes back to the ancient Greeks. The Greek philosopher Democritus thought that matter was made up of indivisible atoms. He was partly right: it is made up of atoms, but they are not indivisible. According to scientists today, it is quarks and some other subatomic particles that are indivisible, although there are different kinds of quarks.

GEOMETRY AND INFINITY: MY CASE AGAINST PI

The quest for an indivisible particle parallels the question in geometry of how small numbers can be, and this brings me to my other passionate challenge: Do we really need pi? Geometry today is still heavily influenced by the writings of Euclid, who believed that a point is infinitely small and has no dimension. In Chapter 11, I challenge this notion and discuss how the idea that a line is composed of infinitely many points of no dimension or width leads to contradictions. One of these contradictions is Zeno's Paradox. The solution I propose is to say that points have area and lines have width. This blocks the need for infinity that occurs as a result of saying that points have no area.

The regress to infinity that occurs in Zeno's Paradox is a result of neglecting the fact that any measurement requires a unit of measurement and a degree of precision. Zeno's Paradox is only possible because the units of measurement keep shifting. As for precision, a measurement can be carried out to one decimal point, one hundred decimal points, or one million or trillion decimal points, provided we have the necessary tools to make the measurement. However, once the precision is specified and made, then the measurement is complete; it is not an approximation of a measurement carried out to infinity. Of course, we can always ask for a more precise measurement, but then the measurement needs to be made again to a higher degree of precision. Like the first measurement, this is not an approximation of a measurement carried out to infinity.

It is the failure to understand that any measurement requires a unit of measurement and a degree of precision that leads to the use of π in the formula for the area of a circle. The formula is doomed from the start since its unit of measurement is the radius squared (r^2). The radius squared is the area of a square with four sides that equal r in length. This is asking how many squares will fit inside a circle. Not surprisingly, the answer is "not a rational number," and so the need for π was born. The number π is an irrational, nonrepeating decimal number that has been carried out to 105 trillion places.

In "The Rope Experiment" in Chapter 11, I address the issue of how to find the area of a circle. I propose taking the circumference of a circle in the form of a rope and laying it flat. When a rope is laid flat, its length can be measured by conventional means with a ruler, a tape measure, or even electronically. What is not needed here is the use of π to describe its length. The measurement of the length of a rope laid flat is a rational number that is made using certain units (e.g., inches, meters, etc.) to a specified degree of precision. This length is not an abbreviation for an infinitely long number like π. Yet the rope doesn't change length when it is laid flat. It is the same length if it is in a circle, laid flat, in the form of a lasso, tied around a tree, or whatever shape it takes. Using this framework, I show that it is possible to derive rational values for the area of a circle as well.

The need for π arises only in the context of the frame of reference of the ratio of the circumference of a circle to the diameter of a circle. Yet according to the rope experiment, this leads to contradictions since the length of the rope, and hence the circumference of the corresponding circle, is actually a rational number. The reason π comes into play is due to a failure to appreciate the need to specify a unit of measurement and a degree of precision when making a measurement. Underlying this is the mistaken idea that a line is made up of infinitely many points of no area. Once we recognize that points have area, the contradictions disappear. It is amazing that scientists and engineers have made incredible technological progress since the Greeks, but we have not advanced our geometry beyond that of Euclid.

PAST, PRESENT, AND FUTURE

I wrote this book while running a market research company focusing on flowmeters fulltime. While it was sometimes difficult to find the time to write the book, these ideas are too important to leave unaddressed. Some of these ideas, especially the mathematical ones, are elaborations of ideas I have had since attending graduate school at Rockefeller University. I plan to continue to write about these and related ideas as time and circumstances allow. I hope you find this book interesting and, even if you don't agree on some points, will find it a worthwhile expression of an alternative point of view.

Jesse Yoder, PhD
Flow Research, Inc.
Wakefield, Massachusetts

About the Author

Jesse Yoder, President of Flow Research, Inc., USA, has 36 years of experience writing about and analyzing process control and instrumentation markets, beginning as President and Founder of Idea Network, USA. He received a B.A. degree in philosophy at the University of Maryland, USA, in 1973 and a PhD in philosophy from the University of Massachusetts, USA in 1984. Dr. Yoder was an adjunct philosophy professor at the University of Massachusetts, Lowell, USA and Lafayette College, USA for 10 years. He founded Flow Research, Inc. in 1998, where he still serves as the President. Dr. Yoder has received two patents for new dual-sensor flowmeter designs. Several prototypes of these designs have been built and tested at CEESI. Dr. Yoder has prepared over 300 market studies, published over 300 articles on flow and instrumentation in industry journals, and has written several books, including *New-Technology Flowmeters* (CRC Press, 2023), *Conventional Flowmeters* (CRC Press, 2023), and *The Tao of Measurement* (2015).

Acknowledgments

First, I would like to acknowledge my appreciation for both flow and philosophy. To me, they are very closely interrelated. Flow is a continuous phenomenon and understanding and analyzing continuity has become something of a life's work and goal for me. Many philosophical puzzles, including those about the number line, are related to the notion of continuity. In this book I explore both concepts together.

I am especially grateful to the many people in the field of flow and instrumentation who have supported my research, and the ongoing projects at Flow Research. I have been privileged to work with some of most brilliant minds in the industry. There are too many to acknowledge them all, but I would especially mention Mike Pins, Lonna Dickenson, Dr. Karin Kettenbach, Mike Touzin, Dan Cychosz, Jennifer Zaugg, Roberto Guazzoni, Jessica Lackey, Jana Andreas, Ellen Degnan, Dick Laan, Juan Carlos Ballesta, Yuichi Kikuchi, Frank Frenzel, Carsten Habersetzer, Uwe Hettrich, Lammert Heijnen, Steve Kannengieszer, Frank Sass, Chris Waller, Dr. Alfred Rieder, Dr. Mark Bell, Dr. Kevin Warner, Paul Warner, Jesse Osborne, and Todd Sullivan. There are literally hundreds or perhaps I should say thousands of people who have helped me along the way since I began doing market research in 1991, and especially since founding Flow Research in 1998. Brian Crosby has been a good friend since the beginning. I would like to thank Vicki Tuck for all her support throughout the years.

Belinda Burum has been a wonderful friend to me for 35 years and has served as a partner in building Flow Research. She has been my source of inspiration during this entire time. I could not have built Flow Research without her. Nicole Riordan and Leslie Buchanan have also stood by me for many years. I am grateful to Taylor & Francis for the opportunity to write this book, and for supporting me in my earlier books. Sonia Tam and Kyra van Nostrand have been extremely helpful in putting this book together.

I am grateful to my philosophy professors at the University of Maryland, The Rockefeller University, and the University of Massachusetts Amherst who taught me how to write and how to analyze concepts. Somehow, my love for philosophy and flow has become joined, and I love to write about both of them. I hope to continue to write for as long as God gives me the strength and capability.

Jesse Yoder

1 Starting Point

WHY WRITE ABOUT MASS FLOW?

There are many ways to divide up flow and flow measurement. Two of the most important ways are as follows:

- Volumetric flowrate
- Mass flowrate

Volumetric flow measurement tells you how much of a fluid (water, liquid, gas, or steam) goes past a point (such as the cross-sectional area of a pipe) in a specific period of time. It is typically expressed in cubic meters per second (m^3/sec), liters per minute (L/min), and cubic feet per second (ft^3/sec). For example, if 20 cubic feet of water flow through a pipe in one second, the volumetric flowrate is 20 ft^3/sec.

Mass flow measurement tells you how much mass moves past a point (such as the cross-sectional area of a pipe) in a specific unit of time. It is typically expressed in kilograms per second (kg/sec) or pounds per minute (lb/min). Other units are grams per minute (g/min) and kilograms per hour (kg/hour). Mass is considered to be the amount of "stuff" or matter in an object or a fluid. This is how we experience mass in the objective world, using our five senses. Mass in the scientific world of atoms and molecules amounts to the sum total of the atomic weights of the atoms and molecules that make up a substance or fluid.

Volumetric flow and mass flow are related through density. Density is the amount of mass of a substance per unit volume. It tells you how much "stuff" (mass) is in a given amount of space.

The formula for density is Density = Mass/Volume. Using the same formula, we have Mass = Density * Volume. Unlike mass, density varies with temperature and pressure, especially for gases.

The relation between density and volume is easily shown by a few examples. Imagine two boxes of equal sizes. One is filled with rocks and the other is filled with feathers. While both boxes have the same volume, the one filled with rocks is denser than the one filled with feathers. Rocks have more "stuff" packed into a small space than feathers do. Rocks are composed of dense minerals, which contribute to their overall mass. Feathers are made of a protein called keratin, which enables feathers to be lightweight but strong, enabling birds to fly. Birds that had rocks for wings would probably not get very far.

Density is influenced by temperature and pressure. When pressure is applied to a substance, it forces the molecules of the substance together, resulting in a smaller volume and a higher density. This is especially true of gas. When a substance is heated, it causes the molecules to scatter. This results in greater volume, but lower density. This effect is also more pronounced for gases than for liquids.

DOI: 10.1201/9781003457855-1

WHAT IS MASS?

Chapter 2 gives some of the basic concepts and definitions surrounding mass flow. These include accuracy, velocity, cross-sectional area, precision, turndown, and other important flow concepts. Because temperature sensors are especially important in multivariable flow, this chapter includes a table on different types of temperature sensors. It also contains a table of common temperature values and a list of useful definitions of units of length.

Chapter 3 is devoted to a discussion of mass: what it is, and how it relates to our ordinary experience and to scientific views of mass. This chapter distinguishes between how we experience mass in our daily lives (objectively physical mass) and the scientific views about mass (scientifically physical mass). Even though we readily experience mass in our daily lives through heavy and light objects, the explanation for our experiences is rooted in atomic theory and in complex views about the nature of protons and neutrons.

This chapter is a detailed look at the concept of mass. It starts with the concept of matter and describes how matter is made up of molecules and atoms. It then looks at the components of atoms, which are protons, neutrons, and electrons. As it turns out, the concept of mass is deeply rooted in the fields of energy created by the subatomic particles that make up protons and neutrons. These include quarks, gluons, and other subatomic particles. It is not so much that mass is inherent in these subatomic particles. Instead, the mass of a proton derives from the fields of energy created by the fact that these particles are moving at near light speed.

At another level, and closer to our level of experience, the mass of a substance is due to the atomic weight of the atoms and molecules of which it is composed. Hydrogen has the lowest atomic weight and is the lightest element. Oxygen has a higher atomic weight and is heavier. These examples show that the mass of a substance is rooted in elements of matter of which we have no direct experience.

MEASURING MASS FLOW

Chapter 4 describes four ways to measure flow:

- Coriolis flowmeters
- Thermal flowmeters
- Mass flow controllers
- Multivariable flowmeters

It describes each of these types of flowmeters and gives their advantages and disadvantages.

Multivariable flowmeters determine volumetric flow, and then use pressure and temperature sensors to calculate density. They then determine mass flow by multiplying volumetric flow by density.

The next chapters are three of the most important in the book. They give an in-depth view of Coriolis flowmeters from three perspectives: historical, theoretical, and current developments. Chapter 5 traces the history of the development of

Coriolis flowmeters from the early 1950s until today. It discusses two patents in 1958 and 1960 that try to use the Coriolis principle to actually rotate the flow. In 1965, Anatole J. Sipin published a patent in which he criticized the idea of rotating the flow and proposed oscillating the tubes instead. This idea stuck, along with the name, and in 1978, Jim Smith's patent for a flowmeter using oscillating tubes was granted.

Chapter 6 takes a close look at Coriolis theory. This chapter takes a look at the case for saying that Coriolis meters rotate the flow, and saying that they oscillate the flow. It argues that oscillation is not a form of rotation. It further presents an alternative view on the theory of Coriolis flowmeters according to which it is the inertia of the fluid that causes the deflection of the flowtubes and the consequent ability of the Coriolis flowmeters to measure flow.

Chapter 7 traces the development of Coriolis flowmeters from their introduction in 1977 through to today. While Coriolis meters were first introduced by Micro Motion in 1977, there is some controversy about which European company was the first to introduce Coriolis meters in Europe. Heinrichs, KROHNE, Endress+Hauser, and Rheonik all have claims to this distinction.

Chapter 8 examines the history and operating principle of thermal flowmeters. It draws on original interviews made with the founders of the original flowmeter companies: Bob Deane, Mac McQueen, and John Olin. John Olin and Jerry Kurz worked together on developing thermal flowmeters in the early 1970s until they obtained an "industrial divorce" in 1977. As a result, Kurz Instruments was the first to bring thermal flowmeters to market. Bob Deane and Mac McQueen founded Fluid Components, Int'l in 1964 as a flow switch company. The company began offering thermal flowmeters in 1981.

Thermal flowmeters measure mass flow by measuring how quickly heat dissipates in a heated flowstream. They do not measure mass flow directly because they need to take gas composition of the mixture they are measuring, along with the heat transfer properties of the gases involved.

Chapter 9 – Mass flow controllers are unique in that they both measure and control the flow. Most have a bypass built into the flowmeter. Part of the flow is diverted into an upside-down U-shaped capillary tube, where the flow measurement is made. After the flow is measured, it rejoins the main flowstream.

The four main technologies used in mass flow controllers are thermal, Coriolis, differential pressure (DP), and ultrasonic. Of these four technologies, thermal is by far the most dominant. Coriolis meters do not have a bypass for measuring the flow.

Mass flow controllers are widely used in the semiconductor and industrial markets. They specialize in low flows, and most are two inches or less in size. In industrial markets, they are used in a wide variety of applications, including life sciences, fuel cells, hydrogen, biogas, heat treating, lab-grown diamonds, and many others.

Chapter 10 is devoted to multivariable flowmeters. There are two main types of flowmeters that can measure several parameters besides volumetric flow to calculate mass flow. These variables are static pressure, temperature, and density. These two main types of flowmeters are as follows:

- DP flowmeters
- Vortex flowmeters

Measuring mass flow is significantly more involved than simply measuring volumetric flow. This is because the mass of a fluid varies with temperature and pressure. It is also dependent on the particular gas being measured, and the molar mass of that gas. To calculate mass flow, the flowmeter or flow computer needs to use as part of its input the molar mass of the gas whose flow is being measured. This means that to adequately determine the mass flow of a gas, the type of gas (e.g., air, oxygen, nitrogen, argon, etc.) has to be input into the flowmeter or flow computer.

One way to accomplish a multivariable measurement is to use a temperature sensor or transmitter along with a pressure sensor or transmitter to measure the temperature and pressure of the fluid whose flow is being measured. The outputs of these temperature and pressure devices are then output to a flow computer. The volumetric flow value is also sent to the flow computer. A flow computer can also store other important values that may be needed to calculate flow. Some of these may be input by the operator. They include the inner diameter of the pipe, the identity of the gas or gas mixture being measured, molar mass values for different gases, the compressibility of the gas (z), humidity, and elevation. If a flow computer is used to determine gas flow, it is making a multivariable measurement. However, this does not make it a multivariable transmitter. If these values are put inside a flow transmitter, and the transmitter makes the flow calculation, then the transmitter is called a multivariable transmitter.

A DP flowmeter consists of a differential pressure transmitter integrated with a primary element and has the capability of calculating flowrate based on difference in pressure. DP flowmeters rely on a constriction placed in the flow line that creates reduced pressure in the line after the constriction. A primary element is used to create the constriction in the flowstream. A DP flowmeter also requires a means to detect the difference in upstream vs. downstream pressure in the flow line. While this can be done with a manometer, today's DP flowmeters use DP transmitters that sense the difference in pressure and then use this value to compute flowrate.

The following types of primary elements are described in this book:

- Orifice Measuring Points
- Pitot Tubes
- Venturis
- Cone Elements
- Flow Nozzles
- Wedge Elements
- Combination Elements (e.g., Nozzle/Pitot Tubes)
- Other (e.g., Dall Tubes, Laminar Flow Elements)

Vortex meters are among the most versatile types of flowmeter. They can readily measure liquid, gas, and steam flow. Other new-technology meters such as ultrasonic and Coriolis can readily measure liquids and gases but have difficulty with steam, and thermal meters are used almost exclusively for gas flow measurement. Magnetic flowmeters excel at measuring liquids but cannot measure steam or gas flow. Vortex meters are the only new-technology meter that readily measures all three fluid types.

Vortex flowmeters are well suited for measuring steam flow, and they are widely used for this purpose. Steam is the most difficult fluid to measure due to the high pressure and high temperature of steam, and because the measurement parameters vary with the type of steam. Main types of steam include wet steam, saturated steam, and superheated steam. Steam is often measured in process plants and for power generation. In addition to their ability to tolerate high process temperatures and pressures, vortex meters have wide rangeability. This allows them to measure steam flow at varying velocities. In process and power plants, steam is often measured coming from a boiler.

FINITE POINT GEOMETRY

Chapter 11 – Groundbreaking Finite Point Geometry, The Rope Experiment, a rational value for pi and for the area of a circle, and an alternative to calculus.

This ground-breaking chapter takes a radical look at mathematical concepts inherent in flow measurement and proposes alternative theories that avoid the contradictions of traditional geometry. We explore geometry from a new perspective and develop the fundamentals of a new geometry called Finite Point Geometry. In this new geometry we show how to derive a rational value for calculating the circumference and area of a circle rather than treating pi (π) as an irrational number that has an infinitely non-repeating decimal. In the process we challenge the father of geometry, Euclid, and his foundational belief that points have location but no extension or area, and we provide a solution to the resulting Zeno's Paradox. In addition, through a new spin on calculus, we offer a fresh approach to tackling one of the most difficult tasks for a traditional geometry based primarily on straight lines: measuring the area under a curve.

We dive deeply into these concepts in the following sections:

1. An overview of geometry as a branch of mathematics
2. Euclid's views on points and the number line
3. Euclid's views on points and lines generate Zeno's Paradox and other paradoxes
4. The prevailing solution to Zeno's Paradox
5. Where Zeno goes wrong
6. Measurement requires a unit of measurement and a level of precision
7. Points lie on the line, not in the line
8. How many points lie on the line
9. From length to area: Measuring the area of a circle
10. Why we need π in the first place
11. The role of π in calculating the area of a circle
12. The Rope Experiment
13. The length of a rope doesn't change when you lay it flat
14. The area of a circle is a rational number
15. Measuring flow in round pipes
16. Application to mass flow
17. Calculus and measuring the area under a curve

We begin by placing geometry in the context of mathematics, then discuss the obvious but important fact that making a measurement requires a unit of measurement and a level of precision. We explore the relationship between points and lines, and propose that points lie *on* the line, not *in* the line.

We look back on how π originated and the various attempts to determine its value. A diagram shows that r^2 is a square and that the formula for the area of a circle, $\pi * r^2$, is essentially finding how many squares will fit into a circular area. We conclude that the answer appears to be no finite number.

The Rope Experiment section then shows that the value of the circumference of a circle is rational. The experiment essentially takes the circumference of a circle and lays it into a flat line, which of course doesn't change the length of the rope. As a flat line, the length can be measured with a ruler or tape measure, and there is no need for π. We can measure the value of π as 3.1416, or we can carry out the value to any level of precision. The value 3.1416 is not the beginning of an infinite series of nonrepeating numbers; it is a rational number that is determined by measurement.

If we substitute C/d for π in the formula for the area of a circle (πr^2), we get $(C/d) * r^2$. Because C is now rational by the rope experiment, and we know that the diameter is rational in most situations, it follows that the area of a circle is a rational number. This is a dramatic change from the traditional interpretation of πr^2, according to which the area of a circle is an irrational number with an infinitely non-repeating decimal.

BIBLIOGRAPHY

Edmund Loy Upp, *Fluid Flow Measurement: A Practical Guide to Accurate Flow Measurement* (Houston, Texas: Gulf Publishing Company, 1993)

2 Fundamental Concepts

FUNDAMENTAL CONCEPTS

Regardless of how flow is measured, there are certain concepts that are important to understand. One of these is the fundamental flow equation:

$$Q = A * v$$

In this equation:

- Q = volume per unit time
- A = cross-sectional area of the pipe
- v = average velocity of the flow

The letter "Q" has traditionally been used to symbolize flow. It was introduced by French scientists in the 19th century. The letter "Q" comes from the French word "Quantité" which means "amount" or Quantity.

ACCURACY

The accuracy of a flowmeter is its ability to provide a volumetric or mass flow reading that corresponds to the volumetric or mass flowrate that is actually passing through the flowmeter at the time the measurement is made. Accuracy is generally stated in terms of percent and is typically stated as a percentage of how close to the actual flowrate the flowmeter reading is. A flowmeter with a 1.0% accuracy level is accurate to within 1% of the actual reading.

CROSS-SECTIONAL AREA

In this case, we are considering flow within a closed pipe. When considering the area of a pipe, it is important to consider the inside diameter of the pipe, since that is the area the flow passes through. The difference between the inside diameter and the outside diameter of a pipe is simply the width of the pipe wall multiplied by 2. So, one way to determine a pipe's inside diameter if the outside diameter is known is to take the outside diameter, measure the width of the pipe wall, and subtract a value that is twice the width of the pipe wall.

Since pipes are round, computing the cross-sectional area involves using the formula for the area of a circle. The cross-sectional area of a pipe is the area of a circle that is seen when looking at the end of the pipe. This area is determined from the inside diameter of the pipe. The inside diameter of a pipe is the distance of a straight line from a point on the inner wall of one side of the pipe through the center of the pipe to an opposite point on the inner wall on the other side of the pipe.

DOI: 10.1201/9781003457855-2

Once the inner diameter of the pipe is known, the cross-sectional area of the pipe can be calculated. The formula for the area of a circle is:

$$\pi * r^2$$

where r is the radius of the inner diameter. Since the radius of the inner pipe diameter is D/2, where D = the inner diameter, the equation for the cross-sectional area of the pipe becomes:

$$A = \pi * \left(\frac{D}{2}\right)^2 = \frac{1}{4}\pi * D^2$$

This means, for example, that the cross-sectional area of a pipe with an inner diameter of 2 inches is equal to π, or 3.14 square inches.

FLOW VELOCITY

Flow velocity is a measurement of how fast the flow is moving per unit of time. A common Imperial (American) unit for measuring flow velocity is feet per second. A common SI (metric) unit for measuring flow velocity is meters per second.

MASS FLOWRATE

Mass flowrate is a measurement of the mass of fluid passing through a specific region of a pipe in a specified period of time. Mass flowrate is often measured in kilograms per minute (kg/min) or pounds per minute (lb/min). The mass of a body measures the amount of material it contains and causes it to have weight in a gravitational field. Mass flowrate varies with changes in temperature and pressure for gases and steam, but changes very little with temperature and pressure changes for liquids.

Mass flow can be calculated from volumetric flow if its temperature and pressure are known. Mass flowrate W can be determined from the volumetric flowrate Q and fluid density ρ:

$$W = Q * \rho$$

Mass flow is especially important in measuring gas flow, since the number of molecules in a gas varies with pressure and temperature. It is less important in measuring liquids because liquids are nearly incompressible. Coriolis flowmeters measure mass flow directly, while multivariable flowmeters measure mass flow indirectly. It can also be said that multivariable flowmeters infer mass flow by measuring volumetric flow and then using pressure and temperature values to compute mass flow. Examples of multivariable flowmeters are multivariable vortex and multivariable differential pressure (DP) meters.

Precision: Precision in flow measurement refers to the number of digits to which a measurement is made. It is not the same as accuracy since it does not imply a correspondence with the actual value of the flowrate in a pipe or open channel.

TABLE 2.1
Relative Advantages of Different Temperature Sensors

Quality	Thermocouples	RTDs	Thermistors	IC Sensors
Range	–400 to 4200°F	–200 to 1475°F	–100 to 500°F	–70 to 300°F
Accuracy	Less accurate than RTD	More accurate than T/C	More accurate than RTD	Highly accurate
Ruggedness	Highly rugged	Sensitive to strain & shock	Less rugged than T/Cs	Sensitive to shock
Linearity	Highly nonlinear	Somewhat nonlinear	Highly nonlinear	Highly linear
Drift	Subject to drift	Less subject to drift than T/C	Less subject to drift than T/C	More subject to drift than RTDs
Cold Junction Compensation	Required	None	None	None
Response	Fast response	Relatively slow response	Faster than RTD	Faster than RTD
Cost	Low cost except for noble metal T/Cs	Higher cost than T/C	Low cost	Low cost
Ease of Use	Application complex due to multiple types	High ease of use	High ease of use within limited parameters	Application can be complex in semiconductor applications

Source: Yoder and Morley (2015a).

Repeatability: The repeatability of a flowmeter is the ability to produce exactly the same measurement under exactly the same conditions. Repeatability is sometimes viewed as being as important as accuracy.

Turndown: The turndown of a flowmeter is the ratio of the maximum flowrate that the flowmeter will measure within its accuracy level to the minimum flowrate that can be measured within the same accuracy level.

VOLUMETRIC FLOWRATE

Q, or volumetric flowrate, is a measurement of the total volume of fluid flowing through a pipe per unit of time. Typical units for volumetric flowrate are liters per second, cubic centimeters per minute, cubic meters per second, cubic meters per hour, cubic feet per second, gallons per hour, and other units.

Temperature sensors are widely used in flow measurement, especially in multivariable flowmeters (Table 2.1).

TEMPERATURE UNITS

Table 2.2 gives conversion formulas for common temperature units.

TABLE 2.2
Common Temperature Unit Conversions

From	To Fahrenheit	To Celsius	To Kelvin
Fahrenheit (degrees F)	°F	(F – 32)*5/9	(F – 32)*5/9 + 273.15
Celsius (degrees C)	(C*9/5) + 32	°C	C + 273.15
Kelvin (K)–	(K – 273.15)*9/5 + 32	K – 273.15	K

Table 2.3 gives common values for different temperature measurement points.

UNITS OF LENGTH: ONE FURLONG AND A DOORWAY

The following are some common units of length measurement that you may encounter in a variety of contexts:

Egyptian cubit – The length of the arm from the elbow to the extended finger (about 3000 B.C.)
Ancient Greek foot – The width of 16 fingers
Ancient Roman inch – The Roman foot divided into 12 sections (the unciae); "Unciae" meaning 1/12 is the plural of "uncia," the origin of the word "inch"
Yard – Decreed by King Henry I in the 12th century, who ordered a standard yard to be constructed with a length of three feet.

The distance between the end of your thumb and the first joint is about one inch. The width of a standard doorway is one yard.

$$1 \text{ meter} = 3.28 \text{ feet}$$

$$1 \text{ inch} = 25.4 \text{ millimeters}$$

TABLE 2.3
Common Temperature Measurement Values

	Fahrenheit	Celsius	Kelvin
Body Temperature	98.6°	37°	310.2
Boiling Point of Water	212°	100°	373
Freezing Point of Water	32°	0°	273
Hot Day	86°	30°	303
Cold Day	35°	2°	275
Room Temperature	68°	20°	293
Surface of the Sun	10,100°	5600°	5900

Source: Yoder and Morley (2015a).

1 foot = 0.3048 meters

1 mile = 5,280 feet

1 mile = 1,760 yards

1 mile = 1.609344 kilometers

1 kilometer = 0.62137 miles

1 yard = 914.4 millimeters

1 yard = 0.9144 meters

1 nautical mile = 1.150779 miles

1 furlong = 660 feet

1 furlong = 7920 inches

1 centimeter = 0.3937 inches

1 inch = 2.54 centimeters

1 centimeter = 10 millimeters

1 meter = 100 centimeters

1 kilometer = 1000 meters

From Jesse Yoder and Richard E. Morley, 2015, The Tao of Measurement, Research Triangle Park, North Carolina; Instrument Society of America

BIBLIOGRAPHY

Jesse Yoder and Richard E. Morley, *The Tao of Measurement* (Research Triangle Park, North Carolina: International Society of Automation, 2015a)

Jesse Yoder and Richard E. Morley, *The Tao of Measurement* (Research Triangle Park, North Carolina: Instrument Society of America, 2015b)

David W. Spitzer, *Industrial Flow Measurement* (Research Triangle Park, North Carolina: Instrument Society of America, 1990

3 A Theory of Mass

When I was an undergraduate philosophy major at the University of Maryland, I learned the phrase "What is mind, no matter; what is matter, never mind." At that time, I took this to be a light-hearted way to point to the duality of mind and body. I later learned that this statement was originally made by George Berkeley, an 18th-century philosopher who argued for idealism. According to Berkeley, the physical world only exists in the mind of God and in the minds of people.

The British philosopher Bertrand Russell told the Atlantic magazine an amusing story about this saying. When Russell became interested in philosophy, members of his family would say to him that all of philosophy can be summed up in the above saying. He recounted that he ceased being amused by the saying after hearing it 60 times, and he was relieved that when he went to Oxford, he found people who didn't think that studying philosophy was absurd (Figure 3.1).

FIGURE 3.1 Bertrand Russell.

DOI: 10.1201/9781003457855-3

Now, as I research flowmeters, I'm finding that understanding "what is matter" – or more precisely, "what is mass" – does matter. I also see how philosophy sheds light on matter as a physical substance. So first, before I delve into mass at the atomic level, I invite you to journey into the realm of Russell and other philosophers who explored the nature of matter and beyond.

IDEALISM AND MATERIALISM

Bertrand Russell, along with G. E. Moore, is widely regarded as one of the founders of analytical philosophy, the predominant method of philosophy taught and practiced in the 20th century in America and Britain, and also so far in the 21st century. Russell's contributions extend beyond philosophy to logic and mathematics as well.

George Berkeley was a contemporary of René Descartes, who is often called the father of modern philosophy. Descartes lived from 1596 to 1650. The span of his life overlapped with that of some other great philosophers of that time, including Thomas Hobbes and John Locke. Besides philosophy, Descartes also made major contributions to both physics and geometry. His name still lives in the Cartesian coordinate system.

While Descartes is known for saying "I think, therefore I am" (Cogito ergo sum), he defined the framework of the discussion in philosophy of mind and materialism with his definitions of mind and body. According to Descartes, body is an "extended thing," while mind or consciousness is an "unextended, immaterial thing." This raises the question of how mind/body interaction can occur. How can an extended thing such as a physical body interact with an unextended, immaterial mind?

Descartes' definitions of mind and body framed the philosophical debate on mind and body for several centuries. One of the arguments Berkeley used for his idealism is that those who believe in the existence of physical things could not show how these objects could interact with a spiritual being. While it is easy to set aside Berkeley's ideas that everything that exists including the physical world only exists in the mind of God or the minds of people as being preposterous, the alternative views are not that much more plausible. Thomas Hobbes followed the Greek philosopher Democritus and argued for materialism. This is the view that everything that exists can be accounted for in terms of material or physical principles. While this view seems more plausible than Berkeley's idealism, it requires accounting for consciousness and perception in purely physical terms.

DESCARTES HAS GIVEN US A FLAWED FRAMEWORK

As stated above, Descartes' framing of the mind/body problem in terms of the relation between "extended things" (body) and unextended things (mind) set the framework for the subsequent debate on this issue for centuries to come. George Berkeley was an idealist (all is mind), while Thomas Hobbes was a materialist (all is body). Spinoza believed that mind and body are two aspects of an underlying substance. In the 20th and 21st centuries, the power of Descartes' framework still exists, although much of the discussion has turned to the relation between mind and brain. For many materialistic and even dualistic philosophers, the term "body" in Descartes'

FIGURE 3.2 René Descartes.

framework has come to mean "brain." Some philosophers attempt to show the identity of mind and brain, or attempt to show that mental phenomena and perception are reducible to brain functions. More recently, philosophers such as Daniel Dennett and David Lewis have taken up this position (Figure 3.2).

ANOMALOUS MONISM

Someone who is inclined to reject both the "all is mind" idealism and the "all is body" materialism is still left with accounting for the interaction of mind and body. In a 1970 paper called "Mental Events," Davidson proposed a theory called "anomalous monism." It is important to understand that Davidson substitutes "physical events" for talk of body, and "mental events" for talk of mind. He defines "physical events" as ones that have a physical description. He defines "mental events" as ones that have a mental description. According to this view, mind and body are not discrete substances, yet mental events such as thoughts are not reducible to physical events, or to events in the brain (neural events). This view has its roots in Spinoza. Spinoza was reacting to the difficulty in accounting for mind/body interaction, using Descartes' seemingly irreconcilable definitions of mind and body. According to Spinoza, mind and body are not separate substances, but instead are two different attributes of a single underlying substance.

Davidson formulates his anomalous monism as follows:

> Anomalous monism resembles materialism in its claim that all events are physical, but rejects the thesis, usually considered essential to materialism, that mental phenomena can be given purely physical explanations. Anomalous monism shows an ontological bias only in that it allows the possibility that not all events are mental, while insisting that all events are physical.

Davidson has described his position as a combination of conceptual dualism and ontological monism. It is conceptual dualism in that he does not believe that mental descriptions can be reduced to physical descriptions, or vice versa. It is ontologically monistic in that he believes that mental events are identical to physical events. This form of monism is materialism: he believes that, ontologically speaking, mental and physical events are physical events, even if they have mental descriptions.

A part of his defense of anomalous monism reads as follows:

1. For every scientific law L there exists a set of concepts and descriptions which forms a closed, homonomic domain whose elements are drawn on by L.

According to Davidson, every scientific law is governed by a closed set of concepts and descriptions that are drawn on by the law. By "a closed set of concepts" he seems to mean an interrelated set of concepts and definitions that all draw on a single field of study or body of knowledge. For example, physical laws all draw on terms and concepts defined within physics, while chemical laws similarly draw on terms and concepts defined within chemistry. Presumably, Davidson would say that the laws of physics and chemistry do not draw on mental terms and concepts. This is the import of the term "closed." It does not allow for the use of terms outside of physical and chemistry in stating the laws of physics and chemistry.

While Donald Davidson's anomalous monism represents an important advance over Spinoza's theory, his discussion suffers from the assumption made by so many philosophers after Descartes: that it is possible to divide the world into two distinct realms: the physical (material) world and the mental world. It is only when we reject this dualistic framework made most prevalent by Descartes that we can begin to understand our world.

WHY DESCARTES' DUALISTIC FRAMEWORK MATTERS FOR THE CONCEPT OF MASS

This book is about mass flow measurement. We are approaching this topic by first looking at the concept of mass, then looking at "mass flow," then looking at "mass flow measurement." According to the Cambridge English Dictionary, the term "mass" means "the amount of matter in any solid object or in any liquid or gas." What is most important in this definition is that mass is "the amount of matter in any solid object." The definition of "matter" in the same dictionary is "physical substance in the universe."

Putting the above definitions together, we can define mass as "the amount of physical substance in a body or in any liquid or gas." What is most important in this definition is the use of the term "physical." When Descartes defined body as "an extended thing," he was referring to the idea that a body occupies space; it is three-dimensional. Today we would say that it has length, width, and height. Of course, mass does not account for the entire realm of the physical. According to Einstein's special theory of relativity:

$$E = \mathrm{mc}^2$$

In this famous equation:

E = energy
m = mass
c = the speed of light

According to this equation, mass and energy are equivalent and can be converted into each other.

DOES "PHYSICAL" HAVE A SINGLE UNIFORM MEANING?

In Donald Davidson's anomalous monism, he seems to be thinking of what is "physical" as meaning "what can be described in the language of chemistry and physics." However, many discussions of mind and body today focus on the relation between mental events or entities, such as thoughts, and neural events or entities that occur in the brain. While they may both be physical in some sense, talk of rocks and star explosions that may occur in physics are very different from the language of neurology. In neurology, we use terms like "synapse," "cerebral cortex," and "central nervous system." And if we bring chemistry into the picture, matter is viewed in terms of atoms and molecules. It seems unlikely that we can use the concepts or the laws of physics and chemistry to account for what occurs in the brain, whether or not that is related in some way to what happens in the mind.

There is a similar issue with respect to biology, sociology, psychology, economics, and other sciences. Biology is the study of living things. The concepts and laws of biology deal with evolution, genetics, cells, the development of living organisms, and many other matters unrelated to chemistry and physics. Likewise, sociology deals with the behavior of people in groups. It is the study of the development, structure, and functioning of human society. Each of these sciences has its own conceptual framework and laws, and it is hard to see how they could be accounted for in terms of the language of chemistry and physics.

The point of these examples is to show that dividing the world into mental and physical objects or things, or into extended and unextended things, as Descartes did, does not begin to account for the complexity of our own world, or of the universe as a whole. In fact, trying to define a "physical object" in purely physical terms can be challenging in itself. For example, a cup is a usually round container, often with a handle, that is intended to be used for drinking. A notebook is also a physical object. However, it is a stack of pages, usually ruled and generally bound, that is used for notetaking, journaling, writing, sketching, and other ways to record thoughts and ideas. Like a cup, it is a physical object, but it is defined according to its intended purpose. And even though cups and notebooks are physical objects, they are defined according to their mental purpose.

A true account of the world, science, and our language involves a pluralistic approach that goes beyond the definitions of mind and body. Biology, sociology, and economics are all attempts to explain different aspects of the world that are different from what mind and body. They all involve terms, definitions, and laws that are not included in the language of physics or chemistry, or in descriptions of the mind. However, because this book is about mass and mass flow, instead of further developing this pluralistic view, I will focus on the physical aspects of the world, and on the explanation of matter as physical substance. In doing so, we will find that even the term "physical" has multiple meanings, and that these meanings have important implications for the definition of mass.

THREE CONCEPTIONS OF WHAT IS PHYSICAL

Even though Davidson's anomalous monism may not completely resolve the question about the relationship between mind and body, we can use his concept of a conceptual domain to come to a better understanding of the meaning of "physical." It will become clear that the term "physical" has at least three different senses or meanings.

This book is about mass flow measurement. We are approaching this topic by first looking at the concept of mass, then looking at "mass flow," then looking at "mass flow measurement." For many years I have written about mass flow, volumetric flow, and mass flowmeters. Yet I have struggled to understand what mass is. I understand the word, and I know how to use it, yet I don't feel that I really understand mass or mass flow. Likewise, I have struggled to understand the theories behind Coriolis and thermal flowmeters. This chapter attempts to understand mass by going beneath the surface of observable objects and finding the origins of mass in the incredible energy of subatomic particles moving at near-light speed. Before we do that, however, we will look at the multiple meanings of the word "physical." Once we do that, we can see how the concept of mass emerges from multiple layers of atomic and subatomic elements that, while not directly observable, are nonetheless real.

While not everyone shares my puzzlement with the concept of mass, or my interest in solving the puzzle of mass, there is a good reason why this concept is so elusive. Mass is a property that we cannot observe directly, even though we can observe its effects. Many physical properties like the color yellow, the height of an object, and the hardness of a rock, we can observe directly. These properties are not hidden from us and there is no need to make an inference about them in order to understand them. Similar properties include the taste of ice cream, the feeling of a cool breeze on a hot summer night, and the presence of a tree in our line of vision.

Mass is not like this; it can't be directly experienced. Instead, mass is a scientific concept that embodies certain scientific theories. As a result, to understand mass we need to understand the scientific theories that underlie the concept of mass (Figure 3.3).

FIGURE 3.3 Democritus and the atom.

The attempt to understand the nature of the physical world and what underlies its existence began at least as early as the ancient Greeks. What are physical objects composed of? Can we answer this question by simply dividing an object in half and then continuing these cuts until we reach some kind of fundamental element? What we find in our experience is that when we cut something in half, and then in half again, there is a limit to how many cuts we can make. For example, if we cut a piece of cheese in half and then in half again, there is a limit to how many cuts we can make until the piece of cheese is so small that it can't be further subdivided. This may have to do with the size of our knife or the type of cheese. However, even if we switch to a different type of cheese or use a smaller and sharper knife, we will eventually reach a point where the cheese simply can't be cut any further.

It must have been this type of thinking that the ancient Greek philosopher Democritus went through when he proposed the existence of an indivisible particle called the atom. Democritus, who lived from 450 BC to 370 BC, was the first person to propose this. Democritus theorized that all matter is composed of tiny, indivisible particles called "atoms" that are constantly moving through empty space, and that the different properties of matter arise from the varying shapes and sizes of these atoms. He believed that atoms are solid, indestructible, and cannot be further divided, with their differences in shape determining the characteristics of the materials they form. The word "atom" originates from the ancient Greek word "atomos," which means "indivisible" or "uncuttable."

Democritus has captured an important intuition. In the knife and cheese example, part of the problem is simply the practical one of finding a fine enough knife to cut cheese into very small particles. But what if we move to what is theoretically possible – is it theoretically possible to cut a piece of a physical substance (matter) indefinitely until is no longer divisible? In other words, is there a smallest indivisible particle?

If we were dealing in the world of mathematics, we might answer that a smallest particle cannot be found. Mathematicians have no qualms about imagining the existence of infinite processes. According to contemporary math, the number line is populated with infinitely many points with no area. Mathematicians talk without hesitation about the existence of infinite sets and of the ability to complete an infinite series. Even the formula for the area of a circle, $\pi \times r^2$, makes use of π, an irrational number with no repeating digits. This formula is widely used in flowmeter circles because most pipes are round and it is necessary to know the area of a pipe to accurately calculate flow in the pipe.

JOHN DALTON'S ATOMIC MODEL

Democritus' view of the atom remained the prevalent one for almost 2,000 years. In 1803, John Dalton proposed the first scientific theory of the atom. John Dalton's atomic model was the first complete scientific theory of the atom. It stated that all matter is composed of indivisible and indestructible particles called atoms (Figures 3.4 and 3.5).

FIGURE 3.4 John Dalton.

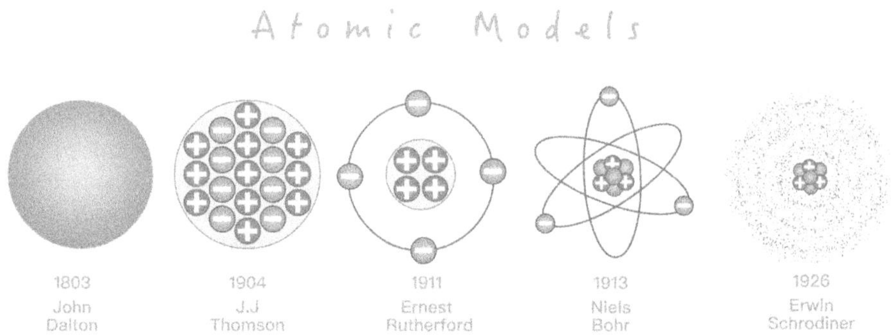

FIGURE 3.5 Representations of different models of the atom.

Dalton's theory included the following key points:

- All matter is made of atoms.
- Atoms of a specific element are identical in mass and properties.
- Atoms of different elements can combine in simple whole-number ratios to form compounds.
- Chemical reactions involve the combination, rearrangement, or separation of atoms.

J. J. THOMSON'S ATOMIC MODEL

The graphic above shows the development of theories of the atom, beginning with John Dalton in 1803. One hundred years later, in 1904, J.J. Thomson proposed a different atomic model. According to Thomson's model, atoms are spheres of positive charge with negatively charged electrons embedded with them, somewhat like plums in a pudding. This model followed Thomson's discovery of electrons in 1897. This discovery resulted from his experiments with cathode ray tubes.

Here are the main points of the Thomson model:

- Atoms are spheres of positive charges.
- Electrons are embedded within these spheres.
- The positive and negative charges are equal in magnitude, making atoms electrically neutral overall.

While Thomson's model lacked empirical evidence to support its claims, it did pave the way for future more accurate models of the atom.

ERNEST RUTHERFORD'S ATOMIC MODEL

The next atomic model, the Rutherford model, was proposed by Ernest Rutherford in 1911. This model describes the atom as having a small, dense, positively charged nucleus at its center, with electrons orbiting around it like planets around the sun. The Rutherford model replaced the earlier "plum pudding" model proposed by

J. J. Thomson, and was a crucial step in understanding atomic structure. Here are some of the main points of the Rutherford model:

- Rutherford's experiments showed that the atom's positive charge and most of its mass are concentrated in a tiny, central region called the nucleus.
- The experiments also revealed that most of the atom is empty space.
- Electrons, with their negative charge, were proposed to orbit the nucleus in a manner somewhat like the way in which planets orbit the sun.

NIELS BOHR'S ATOMIC MODEL

Two years after Rutherford's model, Niels Bohr came out with his own model. According to the Bohr model, proposed in 1913, electrons orbit the atom's nucleus in specific shells, like the way that the planets orbit the sun. Here are some main features of the Bohr model of the atom:

- Electrons move around the nucleus in circular orbits, each corresponding to a specific energy level. Electrons can only occupy certain allowed orbits, called orbital shells.
- Electrons can move between these shells by absorbing or emitting energy in the form of photons. When an electron moves to a higher level, it absorbs energy; when it moves to a lower level, it emits energy.
- The Bohr model improves on the Rutherford model by introducing the idea of specific energy levels that electrons move around in.

ERWIN SCHRÖDINGER'S ATOMIC MODEL

In 1926, Erwin Schrödinger built on the work of Bohr and others, and proposed a different atomic model called the "Quantum-Wave Model." Schrödinger suggested that the movement of electrons in the atom corresponded to wave-particle duality, and consequently electrons could move around the nucleus as standing waves. Here are some of the features of the Schrödinger model:

- It describes the movement of electrons as standing waves.
- Electrons are constantly in motion, and do not have a fixed or defined position within the atom
- The theory does not predict the location of the electron or describe the path it takes within the atom. It only establishes a probability zone to locate the electron. These probability areas are called atomic orbitals.
- These atomic orbitals can be defined between electron clouds and have different energy levels and sub-levels.

CURRENT VIEWS ON ATOMIC THEORY

Today we know that atoms are not a single, indivisible particle, as Democritus thought. Instead, they are made up of protons, neutrons, and electrons. The protons and neutrons form the nucleus of the atom, while the electrons are orbiting

FIGURE 3.6 A representation of an atom.

around the nucleus in constant motion. The term "orbit" suggests that the electron is a solid object traveling around the nucleus, somewhat like the moon orbits the Earth. However, current thinking is the electrons are more like a field of energy than a solid particle, and that they exhibit both wavelike and particle-like qualities. This is the view introduced by Schrödinger, and it is still the prevailing view today (Figure 3.6).

Democritus did not have access to the incredibly sophisticated instruments that we do today such as electron microscopes, X-rays, and laser beams to analyze the contents of material objects or to drill down to the smallest particle. As a philosopher, Democritus conducted a thought experiment using his natural senses and concluded that there must be a fundamental, indivisible particle.

THE WORLD IS NOT AS IT SEEMS

The Latin poet Phaedrus said, "Things are not what they seem; the first appearance deceives many." This ancient saying can be applied to many situations, but it is very fitting to describe the way we experience the physical world and the underlying reality of the world described by physics and chemistry. We experience physical objects as solid, durable, colorful, and as existing over time. We have faith that the dining table we had dinner at last night will still be there in the morning. If the table is made of constantly moving atoms, this is not how we experience it.

WHAT IS MASS?

According to the Cambridge English Dictionary, the term "mass" means "the amount of matter in any solid object or in any liquid or gas." What is most important in this definition is that mass is "the amount of matter in any solid object." The definition of "matter" in the same dictionary is "physical substance in the universe."

Putting the above definitions together, we can define mass as "the amount of physical substance in a body." What is most important in this definition is the use of the term "physical." When Descartes defined body as "an extended thing," he was referring to the idea that a body occupies space; it is three-dimensional. Today we would say that it has length, width, and height.

DOES "PHYSICAL" HAVE A SINGLE UNIFORM MEANING?

The term "physical," at least as we use it in daily life, appears to have multiple meanings. While they may both be physical in some sense, talk of rocks and star explosions that may occur in physics are very different from the term "physical" as used in the language of neurology. And if we bring chemistry into the picture, matter is viewed in terms of atoms and molecules. It seems unlikely that we can use the concepts or the laws of physics and chemistry to account for what occurs in the brain, whether or not that is related in some way to what happens in the mind.

When people talk about their physical appearance, or their physical body, they are not talking about the molecular or chemical structure of their body. They are talking about the physical aspect of their person. Often "physical" in this sense is contrasted with "mind." Philosophers have struggled to explain the relation and interaction between mind and body. The human body has its own unique laws and descriptions that cannot be completely described in the terms of chemistry and physics. A hand moving can be described in terms of muscle contractions and nerve signals, but why it moves (e.g., to wave hello) requires a framework of intentions, goals, and social context – categories that do not exist in physics. Likewise, no equation or law in physics or chemistry can express what it feels like to experience pain, hear a melody, or fall in love. We need biological concepts – like genes, organisms, fitness, adaptation, and development – to make sense of life. These are not merely placeholders for underlying chemistry, but meaningful constructs in their own right.

Even trying to define a "physical object" in purely physical terms can be challenging in itself. For example, a cup is a usually round container, often with a handle, that is intended to be used for drinking. A notebook is also a physical object. However, it is a stack of pages, usually ruled and generally bound, that is used for notetaking, journaling, writing, sketching, and other ways to record thoughts, ideas, and images. Like a cup, it is a physical object, but it is defined according to its intended purpose. And even though cups and notebooks are physical objects, they are defined according to their mental purpose.

THREE CONCEPTIONS OF WHAT IS PHYSICAL

It seems clear that the term "physical" has at least three different senses:

1. Physical properties and entities that exist objectively and can be seen or observed by more than one person at once. These are the physical objects of our daily experience. They are considered physical in the objective sense.
2. Physical properties and entities that can be described in the terms of physics and chemistry. These are considered physical in the scientific sense.
3. Physical properties and entities that make reference to the body of a person, human being, or animal. These are considered physical in the personal sense.

THE OBJECTIVE SENSE OF PHYSICAL

One conception of what is physical involves physical objects that can be perceived at the same time by more than one person. Examples of the use of this sense of "physical" are "I can touch this chair so it must be physical" and "This desk is a physical object." The objective sense of what is physical is the one most commonly experienced in daily life. Objectively physical objects exist independent of any mind. An object that can be seen and touched by one person or by several people is physical in the objective sense. In daily life, we regard physical objects as being solid, stable entities that endure through time.

Attributes of objectively physical objects are not describable in scientific language like physics and chemistry where objects are not really solid, but are masses of atoms, molecules, or waves. In objective reality a desk can be beautiful, round, solid, useful, and made of wood. Other properties such as smoothness, roughness, having a reddish color, weighing more than a computer, and having a sweet taste are examples of properties we ascribe to objectively physical objects based on our perception of them. These properties are described in terms of ordinary language rather than scientific physical language.

There are many natural laws that govern the behavior of objects considered as perceivable objects and not as collections of atoms and molecules. These include the following:

Newton's Laws of Motion:

- First Law (Inertia): An object remains at rest or in uniform motion except insofar as it is acted upon by a net external force.
- Second Law (F = ma): The acceleration of an object is directly proportional to the net force acting on it and inversely proportional to its mass.

Law of Universal Gravitation: Every mass attracts every other mass with a force that is directly proportional to the product of their masses and inversely proportional to the square of the distance between their centers.

Bernoulli's principle: An increase in the speed of a fluid occurs simultaneously with a decrease in pressure or a decrease in the fluid's potential energy.

Laws that govern the behavior of fluids (liquids, steam, and gases) are objectively physical unless the laws describe their behavior at the atomic or molecular level.

Some objectively physical objects are defined in mental or psychological terms. One example discussed earlier is of a cup as a usually round container, often with a handle, that is intended to be used for drinking. It is essential to a racquetball that it be intended to be used to play racquetball. Of course, someone might have other uses for it like playing catch with the dog. Tools such as a hammer and a screwdriver are objectively physical objects designed to be used by human beings for specific purposes.

THE SCIENTIFIC SENSE OF PHYSICAL

Scientific physical objects are ones that can be described in the terms of chemistry or physics. What terms qualify as the terms of physics and chemistry? Some terms of physics or chemistry, such as the term neutron, are introduced originally into the language of these sciences as part of a scientific, physical, or chemical theory. Other terms, such as force, energy, and acceleration, are terms taken from ordinary language, and given a special meaning in a physical or chemical context.

There is no single rule to determine whether a term is a term of physics or chemistry. In chemistry, the best procedure is to consult textbooks of physics and chemistry and the work of physicists and chemists to determine what language they use to formulate their theories. Even the term, "the language of physics," is somewhat misleading because this language has many subsets of smaller languages. However, many examples such as atom, neutron, force, gravity, and mass are clearly terms of physics.

What entities are scientifically physical entities? These are entities that are physical, all of whose properties are scientifically physical. Examples are molecules, atoms, protons, neutrons, electrons, quarks, and gluons. These are all elements of atomic theory.

Other terms such as light wave, radiation, inertia, amplitude, entropy, and supernova are all terms of physics. Terms of chemistry include acid, alloy, element, chemical reaction, atomic mass, electrolyte, ideal gas, nitrogen, radioactive decay, subatomic particle, and viscosity.

THE PERSONAL SENSE OF PHYSICAL

The human body is a living organism, and many of its qualities are designed to preserve life. Someone who doesn't eat well may become sick, while many people work hard to attain their ideal weight and appearance. When we describe someone's personally physical body we use terms like muscular, tanned, blonde, and beautiful. These terms come from the domain of personal discourse.

We also speak of a person's body as having organs such as the heart and liver, and of having perceptual organs such as the ear, eye, and nose. These are biological terms relating to the health and functioning of a human being. Our ears are personally physical objects that enable us to hear, while our eyes enable us to see. Most organs in the human body are designed around specific vital human functions.

HOW THE THREE TYPES OF PHYSICAL PROPERTIES AND ENTITIES RELATE TO EACH OTHER

While it makes sense to think of these as three types of physical properties, they are really three different senses or meanings of the word "physical." One way to characterize the scientific sense is to say that it is the study of non-living matter, or the inorganic world. If we take this view, we could add some other sciences besides physics and chemistry to the list. For example, astronomy, geology, meteorology, and astrophysics all study non-living matter.

It is not implausible to say that some aspects of physics and chemistry deal with physical objects at the observable level rather than at the molecular level. Physics deals with the behavior of inorganic physical objects, their momentum, weight, and mass. It deals with the macroscopic and microscopic behavior of objects. This includes the nature and origin of gravitational, electromagnetic, and nuclear force fields. Physics looks at the origin of the universe, the nature of matter, and the speed of light. Chemistry also looks at chemicals at a macro level. It includes the study of powders, water, hydrogen, copper, and many other elements we experience in our daily lives.

HOW THE MASS OF AN OBJECT IS DEFINED

The British philosopher Bertrand Russell points out the difference between objects as viewed by physics and chemistry and the commonsense view of the world as containing solid objects:

> There is...some conflict between what common sense regards as one thing, and what physics regards as an unchanging collection of particles. To common sense, a human body is one thing, but to science the matter comprising it is continually changing.

In this quote, Bertrand Russell puts his finger on what is confusing about the concept of mass. Mass is a property of physical objects in the objective sense, but it is also a property of physical objects in the scientific sense. Balls, cars, rocks, stars, and trees are all "objectively physical," and they all have mass. However, we cannot see mass; we can only observe its effects. In this respect, mass is like the wind. The mass of an object is determined by the type of molecules making up the object, their individual molecular weights, along with the number of molecules it us made up of. At another level, an object's mass is explained in terms of the "scientifically physical" number and types of atoms it is made up of. The molecular weight of the atoms making up the molecules can be determined by consulting the Periodic Table of Elements.

MASS VS. WEIGHT

Mass and weight are related, though distinct concepts. Mass is the amount of matter or physical substance in an object. This is determined by the type of molecules making up the object, their individual molecular weights, along with the number of molecules it us made up of. The mass of an object is independent of its location, gravity, or shape. An object with a mass of 10 kg on Earth has a mass of 10 kg on the Moon.

Weight is the force experienced by an object due to gravity, and it varies depending on the gravitational field strength. Understanding the difference is crucial in physics, engineering, and everyday contexts, especially when working in environments with varying gravity, such as space exploration.

Mass and weight are related by the formula:

$$W = \mathrm{m} \times \mathrm{g}$$

Here:

W = weight
m = mass
g = gravitational attraction

On Earth, 10 kg weighs 9.81 Newtons, due to the force of Earth's gravity. On the moon, 10 kg would weigh a lot less, due to the reduced gravitational force, even though its mass of 10 kg remains the same.

SIX LEVELS OF WHAT IS SCIENTIFIC PHYSICAL

We can take Russell's world of "continually changing matter" to mean "the world of molecular and atomic motion." There are actually six identifiable levels of what is physical, as shown in Figure 3.7.

Matter is the world of objects as we experience them: solid, observable entities such as rocks, trees, stars, and our human bodies. These are made up of molecules, which in turn are made up of atoms. Atoms are made up of protons, neutrons, and electrons. Protons and neutrons are made up of quarks and other subatomic particles.

Matter is the realm of what is objectively physical. When someone defines mass as "the amount of matter in an object," they are talking about matter as we experience

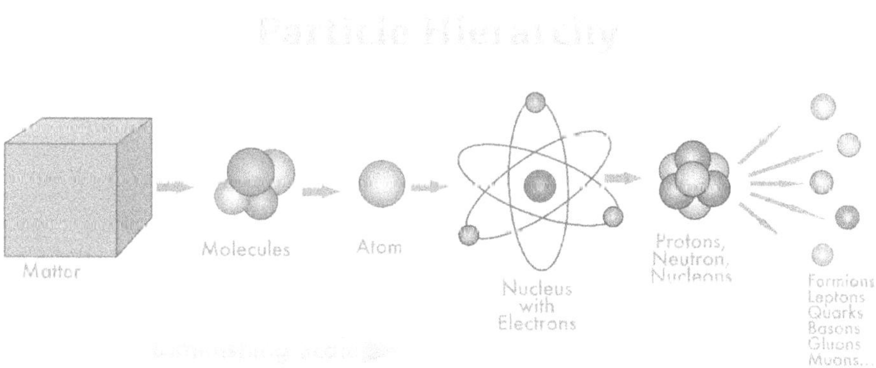

FIGURE 3.7 Six identifiable levels of what is physical.

it with our eyes, ears, and other senses. This was Democritus' starting point when he theorized that objectively experienced matter can be divided down many times over until we reach an indivisible element. We think of objects as being made up of matter. Matter is a very broad concept in that all physical objects are made up of matter.

When Bertrand Russell says that there is a conflict between what common sense regards as one thing and physics regards as an unchanging collection of particles, he is talking about the fact that according to physics, physical objects are made up of molecules and atoms. These molecules and atoms are not static; in fact, they are constantly in motion. Figure 3.7 not only shows how molecules and atoms underlie and make up what we call matter, or material objects, it also shows that molecules and atoms can be further subdivided into a nucleus with protons and neutrons, surrounded by electrons, and from there into further subatomic particles like quarks and gluons.

EMERGENT PROPERTIES

We can explain the relations among these six physical levels through the concept of "emergent property." An emergent property is a property of a system that arises from the interactions and organization of its parts but is not a property of the parts themselves. In other words, an emergent property is a feature or behavior of a whole system that does not exist in any individual component, and cannot be predicted just by analyzing the components in isolation. The emergent property only appears when the parts work together in an organized manner.

Emergent properties are not predictable from the properties of the individual parts. One example of an emergent property is sports fans performing "the wave," You can't understand or perceive "the wave" by studying the behavior of single individual fans. Yet the wave can be clearly perceived when the coordinated actions of groups of fans are observed. Clouds are another example of an emergent property. The shape and color of a cumulus cloud can be observed from the ground; yet the individual water droplets that make up the cloud do not have this color or shape. Fast-moving gas molecules create temperature even though no individual molecule has temperature. Still another example is the roar of the crowd at a football game. The roar of the crowd is the result of the screaming and yelling of hundreds and often thousands of fans. Yet listening to one fan does not yield this roar; it is only the collective action of many fans that generates "the roar of the crowd."

A LADDER OF EMERGENT PROPERTIES

The properties of solid objects we perceive in real life are emergent properties of the molecules and atoms that make them up. Because the atoms are moving so fast, they create the appearance of a solid object. The molecules and atoms themselves are not solid, but when they move at such a high rate of speed, they create the appearance of solidity. A similar thought experiment explains the other levels in the diagram. Molecules are an emergent property of the constituent atoms in motion that make them up, and atoms are an emergent property of the motion of protons and neutrons and the surrounding electrons. Protons and neutrons are an emergent property of quarks, gluons, and other subatomic particles.

The world is not as it seems. What we find when we look beneath the surface is a hierarchy of particles that are built on each other (see diagram above). Today scientists believe that Democritus' indivisible particles are not his atoms, but subatomic particles such as quarks. Beginning at the bottom with subatomic particles, we can build a "ladder of emergent properties" until we reach the stable, solid objects of our everyday experience.

A HIERARCHY OF PARTICLES

In the above discussion, we have drilled down to get to the most elemental concept of mass. If we now build back up, we can say that the much of the mass of protons and neutrons derives from the near light-speed movement of the high-speed quarks and gluons. The mass of an atom derives from its nucleus, which is made up of protons and neutrons. The more protons and neutrons an atom has, the more mass it has. By referring to the Periodic Table of Elements, we can determine how many protons it has. This is its atomic number. For example, hydrogen, with an atomic number of 1, has only one proton. The Periodic Table is ordered according to the atomic number of the elements in the chart. Carbon, with an atomic number of 6, appears to the right of hydrogen, while oxygen, with an atomic number of 8, appears to the right of carbon (Figure 3.8).

While the number of protons in an atom remains constant for the type of element it is, different atoms of the same element can have different numbers of neutrons. For example, Carbon-12 has six protons and six neutrons, while Carbon-14 has six protons and eight neutrons. An atom's mass number represents how many protons and neutrons atoms of this type have. Carbon-12 has six protons and six neutrons and has a mass number of 12. Carbon-14 has six protons and eight neutrons and has a mass number of 14. These different variations of carbon atoms are called isotopes. Like carbon, other atoms also have different isotopes.

The Periodic Table also contains the atomic mass of each element. The atomic mass of each element is the weighted average of the masses of all the isotopes of an element, taking into account their frequency of occurrence. The isotopes of an element have different atomic masses because they have different numbers of neutrons.

MOLAR MASS

Atoms combine to form molecules. The mass of a molecule is the sum of the masses of the atoms it contains. For example, water has the formula H_2O, meaning that one molecule of water contains two hydrogen atoms and one oxygen atom. We can determine the mass of one molecule of water by summing up the masses of two hydrogen atoms and one oxygen atom. However, the molecular weight of molecules may vary, depending on the component isotopes. Oxygen has three stable isotopes, along with several short-lived ones. Oxygen's atomic mass value in the Periodic Table takes these isotopes into account. Oxygen-16 is by far the most common oxygen isotope for oxygen.

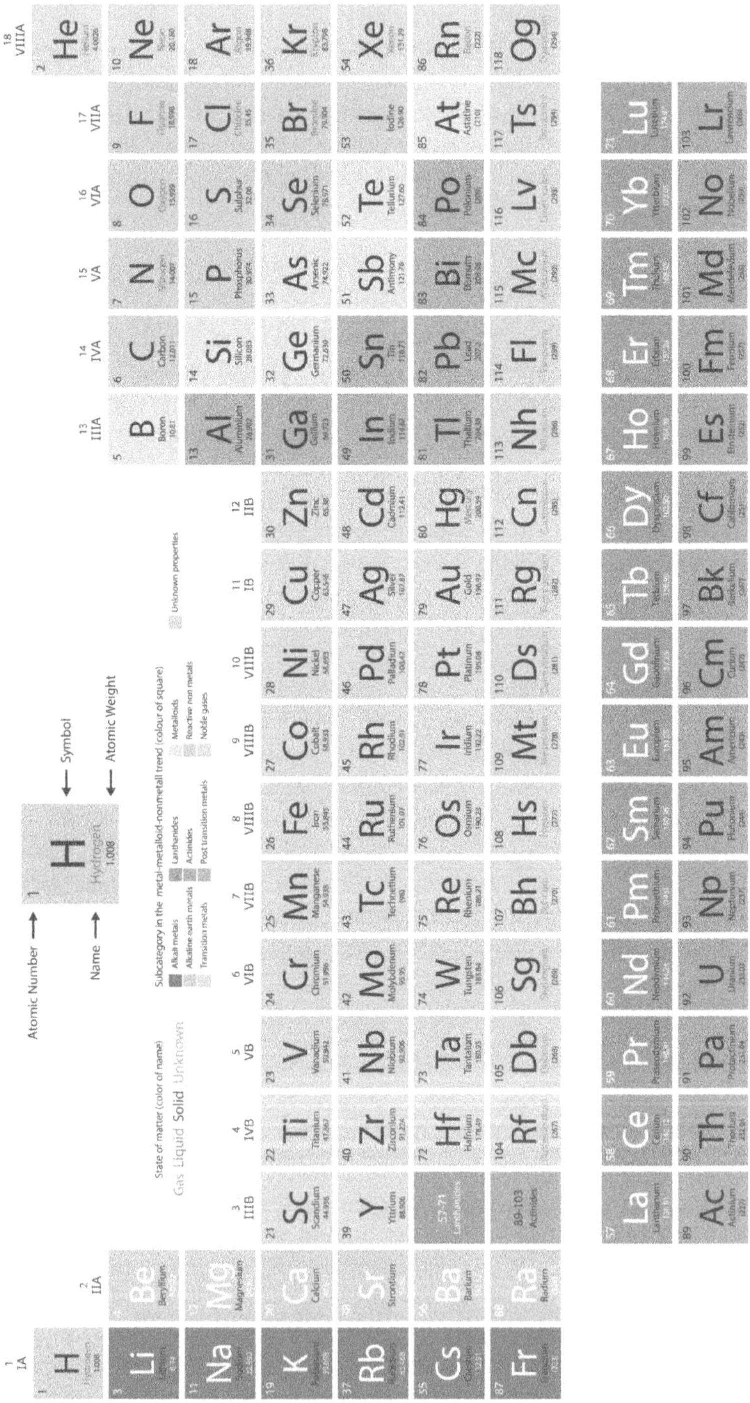

FIGURE 3.8 The periodic table of elements.

While atomic mass represents the mass of atoms, the mass of molecules is called molar mass. The unit of measurement for molar mass is g/mol, meaning grams per mole. A mole is a unit that represents 6.022×10^{23} particles (atoms, molecules, ions, etc.). This number is called Avogadro's number. This huge but important number helps us bridge the gap between the world of molecules and the objective world of our experience.

What is the molar mass of water? We can calculate this by using the atomic mass of hydrogen and the atomic mass of oxygen. A molecule of water has two hydrogen atoms and one oxygen atom. Using the Periodic Table of Elements, we find that the hydrogen's atomic mass is 1.008. Oxygen's atomic mass is 15.999. Therefore, we can compute the using the following formula:

$$(\text{Atomic mass of hydrogen } \times 2) + \text{the atomic mass of oxygen}$$

This works out to be the following formula:

$$\begin{aligned} \text{Hydrogen (H)}\, \text{g/mol} &= 1.008 \times 2 = 2.016 \ \text{g/mol} \\ \text{Oxygen (O)}\, \text{g/mol} &= 15.999 \ \text{g/mol} \\ \text{Total molar mass of } H_2O &= 18.015 \ \text{g/mol} \end{aligned}$$

When calculating molar mass, the common practice is to use the atomic weights listed in the Periodic Table and not be concerned about molecules containing particular or different isotopes. If a particular molecule contains a unique isotope, the molar mass could be calculated by using the atomic mass of that isotope rather than the atomic mass listed in the Periodic Table.

THE MASS OF WATER

What is the value of knowing the molar mass of water? The SI unit for mass flow measurement is kilograms/second. This gives us a way to work down from a mass flow measurement in kilograms per second to the number of molecules being measured. This can be done using the following steps:

1. Convert kilograms/second to grams per second. 1 kg = 1000 grams. For example, if the mass flowrate is 0.25 kg/sec, convert this to 250 g/sec.
2. Convert grams to moles, using the molar mass of water:

$$\text{Moles per second} = (250 \ \text{g/sec})/(18.015 \ \text{g/mol}) = 13.877 \ \text{mol/s}$$

3. Convert moles to molecules, using Avogadro's number:

$$13.877 \ \text{mol/s} \times 6.022 \times 10^{23} = 8.36 \times 10^{24} \ \text{molecules/s}$$

UNDERSTANDING WHAT MASS IS

The above demonstration shows that we can convert any mass flow measurement of water in kilograms per second into a measurement of the number of molecules flowing in a unit time. This can be done for any fluid if we know the molar mass. What are the implications for our understanding of what mass is?

Previous discussion showed that the mass of molecules amounts to the mass of the atoms making up those molecules. Among the 118 elements of the Periodic Table, there is a wide range of atomic mass values. These values are determined by the number of protons and neutrons in the nucleus of the atoms. The heavier atoms have many protons and neutrons, while the lightest one, hydrogen, has only one proton. Electrons and binding energy only play a very small role in an atom's mass.

As discussed previously, protons and neutrons are made up of even smaller particles called quarks. However, the mass of the quarks only accounts for about 1% of the mass of the proton. Most of the proton's mass arises from the interaction between the quarks, gluons, and other particles that are held together by the strong nuclear force. These particles are moving at incredible speeds and the energy they generate is the ultimate source of mass in protons and neutrons. This means that energy is the ultimate source of mass.

What does this tell us about what mass is? The mass of a proton is generated by the energy of the subatomic particles inside the proton. A proton has tremendous energy inside it from the interaction of quarks, gluons, and other subatomic particles. These particles are held together by the strong nuclear force. Yet the proton doesn't have a specific boundary or "skin" holding all these elements in – the internal particles are prevented from "flying out" and are held in place by the strong force that binds them together. This explains why protons, atoms, and even molecules can move at relatively slow paces compared to the near light-speed of the subatomic particles within the proton. Despite all the incredible energy inside the proton, it has a boundary that keeps it together as an entity. Even though this boundary is not well defined, it serves to mark the edge of the proton.

A proton is like a hurricane in many respects. A hurricane is held together by its eye, its eyewall, and by the rainbands of clouds and thunderstorms that spiral outward from the eyewall. Like the proton, a hurricane isn't enclosed in a physical boundary. Instead, the edge of a hurricane is marked by the area where the violent storms that are part of the hurricane dissipate and where the winds fall below hurricane force.

MASS DEFINED

What is mass? Earlier in this chapter, I cited the Cambridge English Dictionary definition of "mass" as "the amount of matter in any solid object or in any liquid or gas." We can now define mass as follows.

From a scientific point of view, mass is determined by the amount of matter in a solid, liquid, or gas. The mass of a substance is measured in molar mass, which is the molecular weight in grams of one mole of that substance. The molar mass of a substance is given by the molecular weight of the molecules that make up the substance in atomic mass units. In the example of water given above, its molar weight of 18.015 g/mol is a result of the atomic weights of hydrogen and oxygen. This molar mass is the sum total of the atomic weights of the atoms and molecules that make it up.

The unit of measurement for molar mass is grams or kilograms. Some substances are heavier than others. For example, hydrogen is the lightest element with an atomic weight of 1.008. Oxygen, by contrast, has an atomic weight of 15.999 g/mol. The differences in mass among different substances are explained by the different atomic weights of their constituents and by how many of those constituent molecules each substance has.

What makes up the protons and neutrons is the fields of energy created by the interactions of quarks, gluons, and other subatomic particles moving at near light-speed and colliding with each other. These subatomic particles are contained in the protons or neutrons of the atoms that make up matter and are held together by the strong nuclear force. The fields of energy that are created by the interactions of subatomic particles are the primary source of mass in the protons and neutrons. The speed with which atoms and molecules move creates the emergent and objective world of matter that we experience as objects in solid, liquid, or gas form.

From an objective (experiential) point of view, we experience physical objects and fluids made of matter. "Matter" is a general term used to designate what makes up any physical object, entity, or fluid. While we don't experience molecules and atoms directly, it is their varying atomic weights and molar masses that account for the differences in the objects we experience. Weight and mass are related but distinct. Objects with greater mass will weigh more because gravity pulls more heavily on their mass.

Another way to describe mass is as inertial mass, which quantifies the resistance of an object to changes in motion. This resistance to change in motion in a solid, liquid, or gas is created by its atomic and molecular mass. The amount of resistance to change in motion is proportional to the atomic weight and molar mass of a substance. Ultimately, the amount of resistance to change is determined by the strength of the field of energy created by the high-speed motion of the subatomic particles in the atomic and molecular structure.

THE BRIDGE BETWEEN THE OBJECTIVE WORLD AND THE SCIENTIFIC WORLD

Earlier in this chapter, we cited the following quote from Bertrand Russell:

> There is...some conflict between what common sense regards as one thing, and what physics regards as an unchanging collection of particles. To common sense, a human body is one thing, but to science the matter comprising it is continually changing.

We can now see that what Russell was referring to as "one thing" is the solid objects in the objectively physical world. When he says that the matter comprising it is constantly changing, he is referring to objects in the scientifically physical world.

This is the world of molecules, atoms, neutrons, electrons, and subatomic particles. This world is a hierarchy of emergent properties. The objects that we experience are emergent properties of the molecules and atoms that make them up. These atoms and molecules are emergent properties of the protons, neutrons, and electrons that make them up. The protons and neutrons are made up of quarks, gluons, and other subatomic particles. When these quarks, gluons, and other subatomic particles move at near light-speed within the protons and neutrons, they generate the emergent property of mass, which becomes the mass of the protons and neutrons. These subatomic particles are held together by the strong nuclear force.

The fields of energy created by the motion of subatomic particles is the primary source of the emergent property of mass in the protons and neutrons and in the atoms and molecules that they make up. These fields of energy are the ultimate source of mass in the object that we perceive in the objectively physical world and that is made up of the atoms and molecules that make up this object in the scientifically physical world.

A BRIDGE BETWEEN MASS FLOW AND THE SCIENTIFIC PHYSICAL WORLD

The demonstration of the way to work down from the unit of mass flow, kilograms/second, to the number of molecules being measured provides a bridge between mass flow as we experience it and the scientific world of molecules and atoms. When we measure mass flow in a fluid, we are measuring its resistance to flow as determined by its molecular structure. Coriolis meters determine mass flow by measuring how much a vibrating tube is deflected by the inertia of a fluid. The force or momentum of the fluid is proportional to its mass, and this determines the amount of deflection, giving us mass flow. The mass of the fluid is determined by the atoms and molecules that make it up in the scientifically physical world.

BIBLIOGRAPHY

Donald Davidson, Essays on Actions and Events (Oxford: Clarendon Press, 1980)

René Descartes, The Philosophical Writings of Descartes, Volume I, Translated by Elizabeth S. Haldane and G.R.T. Ross (Cambridge: Cambridge University Press, 1986)

Bertrand Russell, Our Knowledge of the External World (Abingdon, UK: Routledge Taylor & Francis Group, 2009)

4 Four Ways to Measure Mass Flow

OVERVIEW

There are four main ways to measure mass flow involving four different types of flowmeters:

- Coriolis flowmeters
- Thermal flowmeters
- Mass flow controllers
- Multivariable flowmeters

CORIOLIS FLOWMETERS – THE ONLY FLOWMETERS THAT MEASURE MASS DIRECTLY

Coriolis flowmeters are named after the French mathematician Gustave Coriolis. In 1835, Coriolis showed that an inertial force must be taken into account when describing the motion of bodies in a rotating frame of reference. The Earth is commonly used as an example of the Coriolis effect. Imagine an object thrown from the North or South Pole aiming in a straight line toward a spot on the equator. The object will deviate from the intended straight-line path, resulting in a curved path.

Coriolis flowmeters are composed of one or more vibrating tubes, usually bent. The fluid to be measured passes through the vibrating tubes. The inertia of the fluid causes the tube to deflect, resulting in a twisting motion in the tube. A time difference or phase shift develops between the inlet side of the tube and the outlet side. Position detectors in the form of magnetic pickoff coils sense the positions of the inlet and outlet sides of the tubes. The movement of the pickoff coils generates a sine wave. The time difference or phase shift between the sine waves generated on the inlet and the outlet side of the tubes is directly proportional to the fluid's mass flow. While most Coriolis flowmeter tubes are bent, and a variety of designs are available, some manufacturers have also introduced straight-tube Coriolis flowmeters.

Straight-tube flowmeters work on the same principle as do the bent-tube meters. The inertia of the fluid causes deflection in the first half (inlet side) of the flowtube, and deflection of the flowtube in the second half of the meter (outlet side). This creates a time difference or phase shift in the sine waves caused by the movement of the pickoff coils. This time difference or phase shift is proportional to mass flowrate. Because the oscillatory properties of the measuring tube vary with temperature, temperature is continuously measured so that the necessary adjustments can be made in the measurement.

DOI: 10.1201/9781003457855-4

It is often said that Coriolis flowmeters measure mass flow "directly," unlike other flowmeters that calculate mass flow by using an inferred density value. These other flowmeters measure volumetric flow (Q), which is calculated by multiplying the cross-sectional area of a pipe by the fluid's average velocity. Mass flow is then determined by multiplying volumetric flow (Q) by the density of the fluid. Some multivariable flowmeters measure the pressure and temperature of the process fluid and then use these values to infer fluid density. Mass flow can then be calculated.

Coriolis flowmeters are used on both liquids and gases. While they are highly accurate, they used to be limited in size to six inches or less. Now some Coriolis meters can accommodate line sizes of up to 16 inches. Coriolis flowmeters have relatively high initial cost, although some models are now available in the $3,500 range. Their high initial cost is offset by their normally low maintenance costs. Coriolis meters can handle some fluids with varying densities that cannot be measured by other flowmeters.

Coriolis flowmeters are the only kind of flowmeter that measures mass directly. Many other flowmeters measure velocity or volume; in case a mass flow measurement is needed, it is calculated, using pressure and temperature values. Many of these flowmeters are affected by temperature, pressure, viscosity, and density. By measuring mass flow directly as the fluid passes through the meter, Coriolis meters make a measurement that is virtually independent of changing process variables such as temperature and pressure. As a result, Coriolis meters can be used on a variety of fluids without recalibration and without the need to compensate for process variables that are specific to a particular fluid type. These fluids include most kinds of liquid and gas, and a limited number of them can be used to measure steam flow.

While Coriolis meters have a high initial purchase price and can have high installation costs, end-users today are distinguishing between purchase cost and cost of ownership. Because Coriolis meters have limited maintenance costs and a limited need for recalibration, the life cycle cost of Coriolis meters can be less than competitive meters that have a lower purchase price. Coriolis meters are not subject to wear like orifice plates and some other primary elements. Unlike positive displacement meters, Coriolis meters do not have moving parts that are subject to wear and may require periodic replacement. While the vibrating tube might be considered a moving part, it is made to move by electromagnets, and is not subject to wear under normal conditions. Having low maintenance requirements at a time when many maintenance personnel are retiring or even being laid off is a plus for chemical, life sciences, food & beverage, and other process companies (Table 4.1).

THERMAL FLOWMETERS ARE USED ALMOST ENTIRELY FOR GAS FLOW MEASUREMENT

Thermal flowmeters are used in some of the world's most demanding processes and industries, including chemical, oil & gas, power & energy, water & wastewater, pharmaceutical, pulp & paper, nuclear power, and aerospace.

Although widely used, thermal flowmeters are still not readily accepted as a reliable measuring principle in all industries and markets. Due to the measuring principle itself,

TABLE 4.1
Advantages and Disadvantages of Coriolis Flowmeters

Advantages	Disadvantages
Highest accuracy of any type of flowmeter on liquids	High purchase and installation cost
The only flowmeter type that measures mass directly	Become expensive and unwieldy in line sizes above four inches
Available for line sizes above 6 inches and up to 16 inches	Gas flow measurement can be difficult due to the low density of gas
Can handle sanitary applications with the appropriate materials of construction; Straight tube meters favored for sanitary applications	Pressure drop for bent-tube meters; straight tube meters build to compensate for this
Excel in line sizes of two inches and less and especially in line sizes of ¼ inch or less	Can have a problem measuring liquids with entrained gas
High reliability; low maintenance	Measurement can be affected by vibration
Much new product development ongoing among the major flowmeter suppliers	Have a problem measuring steam flow; Capability limited to dual bent tube meters
Have custody transfer approvals for liquid and gas applications	

they are limited by flow profile dependency, point measurement, a strong dependency on pressure and temperature-related gas characteristics, the negative influence of condensed water, limited sensitivity, and strong temperature dependency in liquid applications.

One limitation of thermal flowmeters is that they are used almost entirely for gas flow measurement. Thermal flowmeters have difficulty in measuring liquid flows because of the slow response time involved in using the thermal principle on liquids. However, some companies now offer thermal flowmeters for liquid flow measurement. In this book, we identify the percent of thermal flowmeter revenues from liquid as opposed to air and gas applications as less than 2%. Clearly thermal flowmeters are used almost entirely for air and gas applications, and applications for liquids play only a minor role in the total market.

A second limitation is in their accuracy. Thermal flowmeters are not nearly as accurate as Coriolis meters, and typical accuracy levels are in the 1% to 3% range. However, thermal suppliers have been improving the accuracy of their flowmeters, with some even claiming as high as ±0.5% accuracy full scale under reference conditions for air measurement. As we discuss in the Frontiers of Research section later in this chapter, suppliers are also making product innovations that improve ease of use and accuracy.

USE IN FLARE GAS MEASUREMENT

Applications in flare gas measurement are a good example of where thermal flowmeters do well in oil & gas. Gas flaring had been commonly used at one time as a method of disposing of excess combustible gases or to relieve the gas pressure within a system. Before climate change became an issue, this method was a widely accepted and economical practice. There were no concerns regarding environmental impacts and no real incentives to maximize the energy content of the gas being flared.

But these reasons have changed since the early 1990s. In today's world, there are numerous incentives to avoid flaring gas, and some rewards for not doing so. Natural gas is a valuable resource, and businesses now seek to maximize its value by recycling excess quantities rather than destroying them. In addition, the pervasive regulatory environment encourages the reduction in emissions of burned gases through sets of both incentives and penalties. Businesses now see the value of reducing their greenhouse gas emissions while acquiring a new source of onsite energy.

The cycle of measuring, recording, and tracking flare and stack gas emissions is likely to intensify over time, and thermal flowmeters – especially multipoint insertion meters – should be a prime beneficiary of this trend.

Another area with good growth prospects appears well downstream within the oil and gas segment: submetering of gas flows. Any large user of natural gas for heating is a potential customer, and the list includes commercial and industrial buildings, campus environments such as hospitals or schools, and multiunit residential complexes. The premise of the application is that energy is a genuine cost center for users of multiple application types, and its accurate measurement should be reliable and fully accountable.

A WIDE RANGE OF APPLICATIONS

The wide range of thermal flowmeter applications includes:

- Industrial and test gas rigs
- Process industries with gas reactions and needs, including chemical, medical, pharmaceutical, oil & gas, food & beverage, automotive, general industrial markets
- Air and gas compressors, compressed air systems, compressed air balancing, boiler air and gas fuel feed lines, natural gas submetering
- Nuclear power plants
- Environmental: flare gas, flare stack monitoring, vent gases, sewage gas, landfill gas, wastewater, aeration flow, digester gas, sewage treatment plants, measuring flue gas from industrial processes, air emission monitoring
- Energy creation: synthesis gas, gasification plants, biogas, biogas in agricultural anaerobic digestion plants
- Energy management, including natural gas and compressed air flow; measuring methane for commercial and industrial applications ranging from chemical production to advanced rocket fuels
- Facilities management, including HVAC and building automation: submetering of gas usage, tenant billing, boiler efficiency, boiler/furnace gas flow, burners and boiler feed control, boiler inlet fuel flow, combustion air, drying air flow, vent air
- Consumption and distribution of compressed air

TABLE 4.2
Advantages and Disadvantages of Thermal Flowmeters

Advantages	Disadvantages
Medium purchase cost but require periodic maintenance	Low to medium accuracy
Ability to measure flow of low-pressure gases	Gas composition must typically be known
Well-suited for flare and stack gas measurement	Can be a safety hazard when measuring flammable gases
Suitable for continuous emissions monitoring (CEM) applications	Presence of moisture or droplets can lead to measurement inaccuracy
Use in wastewater applications to measure air and biogas and to measure sewage gas	Lack of approvals for custody transfer applications from AGA and API
Well-suited for measuring greenhouse gas emissions	Mainly used for gas applications; liquid applications are in the minority
Used to measure emissions from steam boilers, generators, and process heaters	Require periodic recalibration
Insertion models can handle large pipe size measurement	Cannot be used to measure steam flow

- Uncompressed air, nitrogen, and oxygen gas for a wide variety of commercial and industrial applications, from steel production (also argon) to dehydrating fruit to renewing municipal water supplies
- Nitrogen blanketing in storage containers
- Semiconductor gas supply lines, medical oxygen supply facilities, utility gas supply lines in industries, high temperature dry air lines, combustion gas lines, low pressure and low velocity gas lines, incineration facilities lines, electric power plant lines, flowrate measurement systems for automobile testing lines, and other gas supply applications
- Pure gas applications (Table 4.2)

MASS FLOW CONTROLLERS BOTH MEASURE AND CONTROL

Mass flow measurement is the process of determining the amount of mass moving through a cross-section of a pipe per unit time. This measurement is crucial in various industrial processes where precise control and monitoring of fluid flow are required. Unlike volumetric flow measurement, which measures the volume of fluid passing through a system, mass flow measurement accounts for the actual mass of the fluid, making it more accurate for processes where the density of the fluid can vary due to temperature, pressure, or composition changes.

Mass flow measurement, which measures the weight of a fluid independent of pressure or temperature, is more reliable and offers higher accuracy and repeatability than volumetric flow, which measures the volume of a gas or liquid occupying a pipe. This difference is particularly important for gases, which shrink or expand in volume with changes in pressure or temperature, but whose mass – the number of molecules – stays constant.

Fortunately, a mass flowmeter can easily provide flows in volume by adding the volume (the pipe diameter) to the calculation. Conversely, some volumetric flowmeters, for example, multivariable vortex and differential pressure (DP) flowmeters – can indirectly provide mass flow by also measuring temperature and pressure and computing the mass flow.

Flowmeters determine mass flow through mounted temperature, pressure, and flow sensors and this equation: mass flow = (mass density x volumetric flow). They determine density by measuring the pressure and temperature of the flow. They determine velocity by a rotating turbine or vortex sensor.

Strong growth areas in industrial markets for MFCs include the adoption of hydrogen as a clean energy source. Uses of hydrogen include controlling hydrogen gas in fuel cells and electrolyzers. MFCs are used in solar/photovoltaic cell manufacturing processes. MFCs are also essential to battery production, where they are used for controlling gas flows in battery production and energy storage technologies.

The industrial markets are made up of a wide variety of industrial segments. Industrial markets such as automotive are going through a transitional period as uncertainty in consumer tastes becomes more of a challenge to predict. Others, such as emissions monitoring and the life sciences, are thriving due to environmental regulations and upgrades in manufacturing processes. The need to find alternative energy sources will also drive additional research and development in the foreseeable future. MFC manufacturers will benefit from these industrial segments as MFCs are relied upon to measure and control gas flows in these segments. Fuel cells, as one example, represent a more recent growth area for MFCs. This includes the broader category of alternative energy.

Even though MFCs have an advantage over variable area flowmeters and needle valves, this combination of VA meters and needle valves still remains a low-cost solution for some gas flow measurement and control applications. Competition also exists from Coriolis flowmeters and other multivariable flowmeters that measure mass flow. Some Coriolis flowmeters also have a control function, and manufacturers of this MFC design type have claimed an advantage in not requiring a bypass channel to measure flow. While thermal MFCs remain dominant for gas flow measurement and control applications, the presence of other flow technologies is a limiting factor to the dominance of thermal technology.

Mass flow controllers employ a bypass that diverts part of the flow from the main flowstream in an upside-down U shape. They use one of the following four methods to measure mass flow:

- Coriolis
- Thermal
- DP
- Ultrasonic

The advantages and disadvantages of MFCs are discussed in Table 4.3.

TABLE 4.3
Advantages and Disadvantages of Mass Flow Controllers

Advantages	Disadvantages
Very good at measuring low flowrates	Limited to small line sizes; rarely over two inches
Most can both measure and control flow at the same time	Largely limited to measuring gas flow, including very limited application to liquids
Widely used and relied on in the semiconductor industry	Mostly limited to measuring gas flow; limited applications for liquids
Have a small footprint	Cannot be used to measure steam flow
Have a choice of four technologies to measure flow	
Industrial MFCs are well-suited for renewable applications such as fuel cells, solar/photovoltaic, life sciences, and many other applications	
Measure mass flow	

MULTIVARIABLE FLOWMETERS USE MULTIPLE WAYS TO MEASURE FLOW

Multivariable transmitters are almost entirely used for flow measurement, especially for measuring steam and gas flow. Multivariable DP transmitters typically measure volumetric flow by measuring the difference in the pressure of a fluid upstream and downstream of a constriction in the pipe called a primary element. They compute volumetric flow by using this pressure differential which is proportional to flowrate, then use process pressure and/or temperature to calculate mass flow. Multivariable DP transmitters are typically used to measure mass flow of steam and gas. Because the mass of both steam and gas varies with pressure and temperature, it is important to have accurate readings of temperature and pressure to get a correct reading of mass flow.

MULTIVARIABLE DP TRANSMITTERS

Bristol Babcock introduced the first multivariable pressure transmitters in 1992 (the company was acquired by Emerson in 2006). Since that time, other companies have followed suit, including ABB, Honeywell, Foxboro (a Schneider Electric brand since 2014), and Yokogawa. Multivariable transmitters are ones that measure more than one process variable in a single instrument. Typically, these transmitters measure pressure, DP, and temperature. In some cases, they use these values to produce a mass flow measurement.

Multivariable transmitters can reduce the need for multiple devices to perform a single function such as a mass flow measurement output. Originally, in this instance, operators found that they no longer needed to have a number of different transmitters, but that they would need only a multivariable transmitter and a mass flow calculator to perform this flow calculation. A multivariable transmitter would calculate the mass flow as a function of DP, temperature, and absolute pressure, and then output these values. As the solution evolved, manufacturers placed the computing power of a mass flow calculator on board the multivariable transmitter, reducing the extended footprint of the instrumentation as well as performing the calculation in a single device.

The trend toward multivariable transmitters can be expected to continue in the DP transmitter and flowmeter markets. These products typically sell for less than it would cost to buy the transmitters separately and reduce the number of devices to be verified as part of any process quality maintenance routines, further reducing long-term maintenance costs. And besides offering a more economical way to measure flow, multivariable transmitters can measure mass flow as well as volumetric flow. This makes them useful for measuring steam and gas, in addition to liquids. While they do not achieve the same level of accuracy as Coriolis flowmeters, not all applications require Coriolis-level accuracy.

Multivariable DP flowmeters measure more than one process variable, usually DP and/or pressure and temperature. Many of them incorporate a pressure and a temperature transmitter into a single device, making it unnecessary to order these products separately. Multivariable DP transmitters are mainly used to measure mass flow, and thus they can be used for steam and gas flow measurement.

Multivariable DP transmitters require a primary element to create the restriction in the flowstream that enables the DP flow measurement. They can be used with orifice plates, averaging Pitot tubes, Venturi tubes, and other primary elements. They are increasingly popular as end-users seek to reduce costs, and also to do more flow measurement of steam and gas. Another attractive feature of these instruments is that they also provide more dynamic information about the process, enhancing management control and improving quality outcomes.

MULTIVARIABLE VORTEX FLOWMETERS

Multivariable flowmeters measure more than one process variable. In addition to process fluid flow, multivariable vortex flowmeters have one of the following three sensor configurations:

- Integrated temperature sensor; no pressure sensor
- Integrated temperature and pressure sensors
- Integrated pressure sensor; no temperature sensor

Vortex flowmeters display a growing trend towards multi-technology flowmeters. Single variable vortex meters measure volumetric flow. Multivariable vortex flowmeters include pressure and/or temperature sensors for the purpose of measuring mass flow. By using the information from these sensors, the flowmeter can determine volumetric flow, temperature, pressure, fluid density, and mass flow. This multivariable flowmeter is one of a growing number of multivariable new-technology flowmeters, including multivariable magnetic flowmeters and multivariable DP flowmeters. Multivariable ultrasonic flowmeters are popular in district heating applications.

The primary advantage of the multivariable flowmeters is that they can measure mass flow. They do so by measuring the volumetric flowrate as well as either the process flow temperature and/or pressure. Mass measurements are often a preferred option as they reduce the uncertainties inherent to volumetric measurements which – by themselves – cannot fully account for pressure and/or temperature changes in the fluid flow. This can be problematic when the meter is used in gas flow applications. Multivariable vortex meters are especially adept at measuring steam flow.

MULTIVARIABLE ULTRASONIC FLOWMETERS

Ultrasonic flowmeters, including both transit time and Doppler, measure the velocity of a fluid using sound waves. From this they compute volumetric flowrate (Q) using the following formula:

$$Q = v \times A$$

Here:

- v = average fluid velocity
- A = cross-sectional area of the pipe

The output from this calculation is volumetric flow in units like m^3/s or gallons/min.

To get mass flow ($\dot{m}$) from an ultrasonic flowmeter, you multiply the volumetric flow times the fluid density:

$$\dot{m} = \rho \times Q$$

Here:

$\dot{m}$ = mass flowrate (kg/s)
ρ = fluid density
Q = volumetric flowrate (m^3/s)

Density is not measured by the ultrasonic meter itself. To get the density measurement, you need:

- An external density measurement, such as with a densitometer, or
- A calculated density based on pressure, temperature, and fluid composition (especially for gases)

In many cases, these values, including the densitometer values, are fed into a flow computer, which calculates the mass flow using the volumetric flowrate. In other cases, these calculations are done within the transmitter of the ultrasonic flowmeter.

TURBINE FLOWMETERS

Like ultrasonic flowmeters, turbine meters do not measure mass directly. Instead, they measure volumetric flow and multiply that value times density to determine mass flow. As fluid flows through the turbine meter, it causes the turbine blades to turn. Flowrate is proportional to the speed with which the rotors are turning. A magnetic or optical pickup detects the speed of the rotor. Volumetric flowrate (e.g., in m^3/h) is calculated from the rotor speed.

The output from this calculation is volumetric flow in units like m^3/s or gallons/min.

To get mass flow from a turbine flowmeter, you multiply the volumetric flow times the fluid density ($\dot{m}$):

$$\dot{m} = \rho \times Q$$

Here:

$\dot{m}$ = mass flowrate (kg/s)
ρ = fluid density
Q = volumetric flowrate (m^3/s)

Density is not measured by the turbine meter itself. To get the density measurement, you need:

- An external density measurement, such as with a densitometer, or
- A calculated density based on pressure, temperature, and fluid composition (especially for gases)

In many cases, these values, including the densitometer values, are fed into a flow computer, which calculates the mass by multiplying the volumetric flowrate x density. In other cases, these calculations are done within the transmitter of the turbine flowmeter.

Coriolis are the only flowmeters that measure mass directly. To measure flow directly is to measure it without reliance on other parameters or variables that are used in computing the mass flow value. Flowmeters such as DP, vortex, ultrasonic, and turbine measure volumetric flow, then rely on pressure, temperature, and molar mass to compute density. They derive mass flow by multiplying volumetric flow times density.

Thermal flowmeters do not rely on as many variables as DP, vortex, and others, but they cannot be said to measure mass flow directly. Thermal flowmeters infer mass flow by measuring heat loss to the gas. In order to do this, they need to use the specific heat and thermal conductivity of the gas being measured. Thermal meters need to be calibrated to a specific gas or mixture. While they do not require pressure or temperature compensation like volumetric meters, they do rely on known gas properties – meaning they do not directly measure mass flow.

Coriolis flowmeters rely on the inertial mass of fluid to deflect a vibrating tube. They measure the amount of this deflection on the inlet and outlet side of the pipe. The difference between these two values is directly proportional to mass flow. Because they measure mass flow without relying on other parameters or values, they are the only type of flowmeter that measures mass flow directly.

MOLAR MASS AND DENSITY

In Chapter 3 we discussed the concept of molar mass. Here is the concept of molar mass:

> While atomic mass represents the mass of atoms, the mass of molecules is called molar mass. The unit of measurement for molar mass is g/mol, meaning grams per mole. A mole is a unit that represents 6.022×10^{23} particles (atoms, molecules, ions, etc.). This number is called Avogadro's number. This huge but important number helps us bridge the gap between the world of molecules and the objective world of our experience.

The concept of molar mass is an important one for mass flow. This is because it plays an important role in computing density. What is common to the multivariable flowmeters discussed in this chapter and in this book is that they derive volumetric flow and then multiply this value by density to arrive at mass flow.

As described in the section above on ultrasonic flowmeters:

To get mass flow from a volumetric flowmeter, you multiply the volumetric flow times the fluid density ($\dot{m}$):

$$\dot{m} = \rho \times Q$$

Here:

$\dot{m}$ = mass flowrate (kg/s)
ρ = fluid density
Q = volumetric flowrate (m^3/s)

To get the density measurement, you need:

- An external density measurement, such as with a densitometer, or
- A calculated density based on pressure, temperature, and fluid composition (especially for gases)

There are several ways to obtain the calculated density that is the key to determining mass flow. One way is to use a temperature sensor and a pressure sensor and feed these values into a flow computer, along with volumetric flow. The flow computer uses these values to calculate the density of the fluid being measured. It then multiplies the volumetric flow times the density to derive mass flow.

Using a flow computer to obtain a multivariable reading does not make the flowmeter a multivariable flowmeter. A multivariable flowmeter incorporates a temperature sensor and a pressure sensor, and does the density flow computation itself. It then multiplies the density value times the volumetric flowrate to arrive at mass flow. The two most common types of multivariable flowmeters are DP and vortex.

Molar mass is one of the most important concepts to multivariable measurement, because it is vital to calculating density, especially when measuring gas flow. The mass of gas varies with the type of gas, and with its temperature and pressure. Increasing the temperature of a gas reduces its density, while increasing the pressure on a gas increases its density. The density of a gas is all about how tightly the atoms and molecules are packed together in a given volume. The more tightly gas

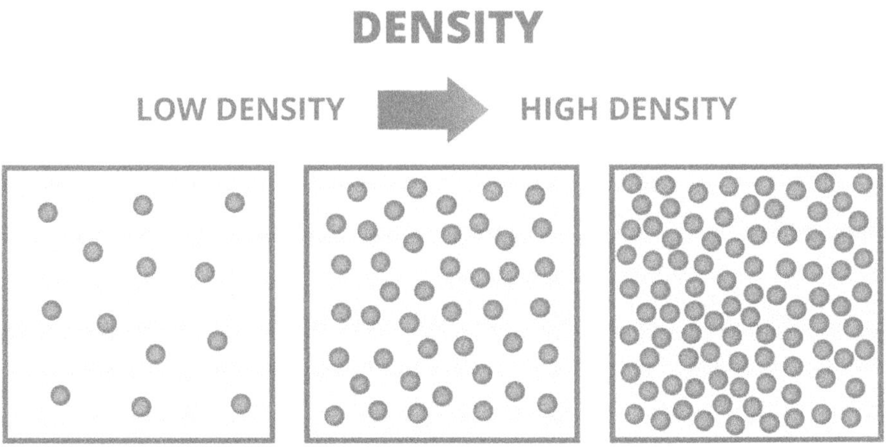

FIGURE 4.1 An illustration of ranges of density.

molecules are packed together, the denser the gas. Popping a balloon is a good way to make the air inside the balloon less dense, because it is suddenly spread around a much larger space. Spraying perfume into a room also makes the molecules less dense because instead of being concentrated in a bottle, they are now spread around the room (Figure 4.1).

The molar mass of a gas reveals how many grams of the gas are in one mole of the gas. As was discussed in Chapter 3, some gases are light, while others are heavier. Hydrogen is the lightest gas, while carbon, nitrogen, and oxygen are heavier. These are individual atoms; the molar mass of a molecule is determined by the molar mass of its atoms and molecules.

In order for a flow computer or flowmeter to compute density, it uses the values of pressure, temperature, and molar mass. This means that it must be told the type or composition of the gas being measured, since molar mass varies with the type of gas. The following formula, which is derived from the ideal gas law, shows the relation between density, pressure, temperature, and molar mass:

$$\rho = (P \times M)/(R \times T)$$

Where:

P = Absolute Pressure (Pa)
M = Molar Mass of gas (kg/mol)
R = 8.314 J/mol · K (R is a constant that relates pressure, temperature, and the amount of gas)
T = Absolute Temperature (K)

This is why multivariable flow measurement is called "indirect," because it is calculated based on the values of these values.

The following shows an example of how to convert volumetric flow to mass flow using molar mass.

How to convert volumetric flow to mass flow using molar mass

1. Start with Volumetric Flow (Q):

$$\text{Example:} Q = 10\,\text{ft}^3/\text{s}$$

2. Convert Units if Needed:

$$1\,\text{ft}^3 = 0.0283\ \text{m}^3$$

3. Calculate Density Using Ideal Gas Law:

$$\rho = (P \times M)/(R \times T)$$

Where:

P = Absolute Pressure (Pa)
M = Molar Mass of gas (kg/mol)
R = 8.314 J/mol · K
T = Absolute Temperature (K)

4. Multiply to Get Mass Flowrate:

$$\text{Mass Flow } (\dot{m}) = Q \times \rho$$

Example:

$$Q = 10\ \text{ft}^3/\text{s} = 0.283\ \text{m}^3/\text{s}$$

$$M = 0.029\ \text{kg/mol (air)}$$

$$P = 101325\ \text{Pa}, T = 298\ \text{K}$$

$$\rho = (101325 \times 0.029)/(8.314 \times 298) = 1.18\ \text{kg/m}^3$$

$$\dot{m} = 0.283 \times 1.18 = 0.334\ \text{kg/s}$$

BIBLIOGRAPHY

Flow Research, Inc., *The World Market for Mass Flow Controllers*, 4th Edition (Wakefield, Massachusetts: Flow Research, Inc., 2025)

5 Gaspard-Gustave de Coriolis and the Early Development of Coriolis Flowmeters

OVERVIEW

Coriolis flowmeters have the distinction of being the only type of flowmeter that measures mass directly. Other flowmeters such as thermal, mass flow controllers, and multivariable flowmeters either measure mass by finding volumetric flow and multiplying this value by density, or by relying on other indirect values. The only exception is that some mass flow controllers employ the Coriolis principle.

Coriolis flowmeters can measure both liquids and gases. However, Coriolis flowmeters have an easier time with liquids. This is because liquids are denser than gases, and Coriolis meters rely on the momentum of the fluid to deflect the flowtube. Gases are lighter than liquids and have a more difficult time deflecting the flowtube. While suppliers have had good success with straight-tube meters for liquids, applying them to gas flow measurement has been more challenging. One major application area for Coriolis flowmeters is downstream applications for petroleum liquids. Here they are displacing positive displacement meters in some applications. Unlike positive displacement meters, Coriolis meters do not have moving parts, apart from the vibrating tube.

This chapter examines the history and development of Coriolis flowmeters and their principle of operation, including illustrations to bring the operating principle to life. Coriolis flowmeters were first introduced to the market in 1977 by Micro Motion. Prior to this was a series of patents that described somewhat similar devices, though none of the products described there were brought to market.

IT ALL STARTED WITH GASPARD-GUSTAVE DE CORIOLIS

Gaspard-Gustave de Coriolis (1792–1843) was a French mathematician, physicist, and engineer best known for describing the inertial forces in rotating frames of reference, now called the Coriolis effect. His work bridged theoretical mechanics and practical engineering, laying the foundation for many modern applications in meteorology, oceanography, and fluid dynamics (Figure 5.1).

 DOI: 10.1201/9781003457855-5

FIGURE 5.1 Gaspard-Gustave de Coriolis.

Born: May 21, 1792, in Paris, France; Died: September 19, 1843 (age 51)
Education:

- Ecole Polytechnique (1808).
- Ecole des Ponts et Chaussees (engineering training).

Career:

- Engineer and instructor at Ecole des Ponts et Chaussees.
- Lecturer and later Director of Studies at Ecole Polytechnique.
- Elected to the French Academie des Sciences in 1836

SCIENTIFIC LEGACY

- The Coriolis effect explains the deflection of moving objects in rotating systems, most famously the rotation of storms on Earth.
- His equations help describe fluid motion on rotating planets, critical in meteorology, ballistics, and engineering.
- Although his work was originally focused on rotating machinery, it now underpins large-scale systems from wind patterns to ocean currents.

CONTRIBUTIONS TO MECHANICS

- Coriolis extended Newtonian mechanics to rotating frames, developed the mathematical description of fictitious forces, and helped formalize the energy concepts used in industrial mechanics. His legacy continues in physics, engineering, and Earth sciences.

MAJOR CONTRIBUTIONS

1. Mechanical Work (1829)
   - Earlier paper: "Du calcul de l'effet des machines"
   - Formalized the definition of work as force × distance
   - Helped establish early energy-based analysis in mechanical systems
2. Coriolis Force (1835)
   - Paper: "Sur les equations du mouvement relatif des systemes de corps"
   - Described how Newton's laws must be modified in rotating frames of reference
   - Introduced the now-famous Coriolis term: $F = -2m(w \times v)$
   - Originally a theoretical mechanics paper, not focused on weather or fluids

Legacy:

- Widely known today for the Coriolis effect in physics and meteorology.
- Namesake of Coriolis mass flowmeters, used in modern instrumentation.
- Influential in multiple fields: fluid mechanics, geophysics, and engineering.

Interesting Fact:

- Although often associated with waterwheels, Coriolis' 1835 paper makes no mention of them. The connection came later through reinterpretation of his work.

Summary:

- Coriolis was a foundational thinker in rotating-frame mechanics whose mathematical insights still power modern technologies, from weather forecasting to precision flow measurement (Figure 5.2).

GUSTAVE DE CORIOLIS' 1835 ARTICLE: "SUR LES EQUATIONS DU MOUVEMENT RELATIF DES SYSTEMES DE CORPS"

This is Coriolis' seminal work in which he attempts to account for movement in a rotating frame of reference. This article is often referred to in discussions of the Coriolis principle. Coriolis wrote this article in French, but as far as we know, it has never been translated into English. For the purposes of research for this book, Flow Research has translated this article into English. The article is quoted in this chapter. Both the original article and the English translation appear on the Flow Research website: www.worldflow.com. Flow Research (Belinda Burum and Jesse Yoder) did the translation.

FIGURE 5.2 Coriolis is one of 72 French scientists and engineers whose names are engraved on the Eiffel Tower.

Source: Photo by Belinda Burum of Flow Research.

In his 1835 article, Coriolis says that in order to account for the motion of an object in a rotating frame of reference, it is necessary to add two supplementary forces. He calls these supplementary forces *compound centrifugal forces*:

> I give this general proposition, namely: that to establish any equation of relative motion of any system of bodies or of any machine, it is sufficient to add to the existing forces two species of supplementary forces; the first ones are always those which must be taken into account for the equation of the living forces, that is to say, they are forces opposed to those which are capable of maintaining the material points invariably linked to the moving planes: the second ones are directed perpendicularly to the relative velocities and to the axis of rotation of the moving planes; they are equal to the double of the product of the angular velocity of the moving planes multiplied by the quantity of relative motion projected on a plane perpendicular to this axis.

These latter forces have the greatest analogy with the ordinary centrifugal forces.

To highlight this analogy, it is sufficient to note that the centrifugal force is equal to the momentum multiplied by the angular velocity of the tangent to the curve described, and that it is directed perpendicularly to the velocity and in the osculating plane, i.e., also perpendicularly to the axis of rotation of the tangent. Thus, in order to pass from these ordinary centrifugal forces to the second forces whose doubles enter into the preceding statement; one has only to replace the angular velocity of the tangent by that of the moving planes, and to substitute for the direction of the axis of rotation of this tangent, the direction of the axis of rotation of these same moving planes. In other words, it is sufficient to substitute for everything that relates in magnitude and direction to the rotation of the tangent, that which relates to that of the moving planes, and to take the double of the forces thus obtained.

It is because of this analogy that I thought I should give these new forces the name of *compound centrifugal forces*: they participate in relative motion by the quantity of motion, and in the motion of moving planes by the use of their axis of rotation and their angular velocity.

We will therefore say that in order to pose an equation of relative motion, which is not that of the living forces, it is necessary to introduce, in addition to this equation, the doubles of *the compound centrifugal forces.*

(see www.worldflow.com for the full text)

Here Coriolis argues that in order to account for the motion of an object in a rotating frame of reference, it is necessary to add two additional fictional forces. These are called *compound centrifugal forces* and they are what later become known as the Coriolis force.

HISTORICAL AND MODERN TERMS: CENTRIFUGAL AND CORIOLIS FORCES

In his 1835 article, Gaspard-Gustave de Coriolis introduced terms and concepts that would eventually form the foundation of modern rotating-frame mechanics. While he did not use the term "Coriolis force," he described it mathematically as part of what he called "compound centrifugal force" (force centrifuge composée).

KEY CONCEPTS

- The "simple" centrifugal force applies to stationary objects in a rotating frame.
- The "compound" centrifugal force (now Coriolis force) acts only on moving objects in that frame.
- The term "double" refers to the factor of 2 required for correctly expressing the Coriolis force.
- Coriolis was focused on rotating machinery, not Earth systems – but the math applies universally.

MODERN PHYSICS SUMMARY

In a rotating frame with angular velocity Ω, the two key fictitious forces are:

$$\text{Centrifugal: } F = m\Omega \times (\Omega \times r)$$

$$\text{Coriolis: } F = -2m\Omega \times v'$$

These terms ensure that Newton's laws remain valid when observed from non-inertial rotating frames.

Coriolis's use of the term "living forces" may seem confusing, and understanding this requires placing the use of the term in the context of the time that Coriolis was writing.

FROM FORCE VIVE TO KINETIC ENERGY

HISTORICAL EVOLUTION OF "FORCE VIVE"

In early classical mechanics, the term "force vive" (French for "living force") was used to describe the quantity of motion associated with an object's velocity. Introduced by Leibniz in the 17th century, it challenged Descartes' idea that momentum (mv) was the true measure of motion.

Leibniz proposed that mv^2, not mv, better represented a body's capacity to do work, especially in collisions and dynamic systems. This idea was widely used in continental Europe and became central to mechanical engineering during the 18th and early 19th centuries.

CORIOLIS AND "FORCE VIVE"

Gaspard-Gustave de Coriolis used the term "force vive" in his 1835 paper to describe the energy associated with bodies moving in rotating systems. His work clarified how this energy transformed in non-inertial frames. He also helped define the concept of mechanical work ("travail") and connected it to force vive.

Over time, the understanding of force vive matured into the modern concept of kinetic energy:

$$\text{Kinetic Energy} = (1/2)mv^2$$

THE WORK-ENERGY THEOREM

Coriolis' work on rotating systems contributed to formalizing the relationship between work and motion. The work-energy theorem states:

The net work done on an object is equal to the change in its kinetic energy.

This theorem bridges the concepts of force (from Newton's second law) and energy (from force vive), and it underpins much of modern physics and engineering mechanics.

It is important to understand that Coriolis was talking about the motion of mechanical objects within a rotating frame of reference. While this would apply to waterwheels, nowhere in this paper does Coriolis mention waterwheels. The idea that Coriolis' article was about the motion of waterwheels came many years later in an effort to explain the intent of the article. Also, the idea that it applies to the rotation of the Earth and to meteorological systems came quite a while after Coriolis' paper was published.

The first person to seriously apply Coriolis' equations to large-scale geophysical phenomena like atmospheric motion was William Ferrel, an American meteorologist, in the 1850s. The following is a historical timeline that shows the broader application of the term to meteorological applications after Coriolis' paper was published in 1835.

FROM CORIOLIS' 1835 PAPER TO METEOROLOGY

This chronicle illustrates the historical development of the concept now known as the Coriolis force, from its mathematical origin in 1835 to its application in meteorology and planetary science.

- 1835 – Gaspard-Gustave de Coriolis publishes his foundational paper on motion in rotating systems, focusing on rotating machinery. He introduces what he calls 'compound centrifugal force'–later known as the Coriolis force.
- 1856 – William Ferrel becomes the first to apply Coriolis' equations to atmospheric motion, explaining why winds and air currents deflect in different hemispheres due to Earth's rotation.
- 1877 – Félix Tisserand incorporates Coriolis terms in celestial mechanics, helping formalize their role in planetary motion.
- 1901 – The term "Coriolis effect" appears widely in meteorological literature, especially in studies of cyclones, ocean currents, and global circulation patterns.

A REVIEW OF EARLY CORIOLIS PATENTS

There were a number of patents filed in the 1950s and 1960s for Coriolis or Coriolis-like flowmeters. Even earlier, in 1917, Ernest F. Fisher on behalf of General Electric Company filed a patent called "Fluid Flow-Indicating Mechanism." As part of this patent, it is claimed:

> It is an established principle that vibrations will be set up in fluid flowing past a statuary object and that the periodicity of the vibrations will be proportional to the velocity of flow of the fluid.

While this sounds somewhat like the formula for a vortex flowmeter, and it attempts to measure velocity rather than mass flow, it does introduce the idea of vibration of the fluid.

In a patent approved in July 1952, Paul Kollsman presented an "Apparatus for Measuring Weight Flow of Fluids." While it does not mention a Coriolis force, it does introduce the idea of measuring the weight or mass of a fluid. To quote from the patent:

> The present invention provides an apparatus for measuring the flow of fluids and provides, more particularly, an apparatus for determining the actual weight flow of a fluid which may be a gas, a liquid, or a mixture of both, the measurement being accurate within a wide range of density and viscosity of the fluid, and uninfluenced by the pressure of the fluid.

The following year, in January 1953, John M. Pearson, on behalf of the Sun Oil Company, patented a meter using a gyroscopic principle. This patent is simply called "FLOWMETER," and the meter was designed to measure mass flow:

> This invention relates to a flow meter designed to measure flow in terms of mass of the fluid.

While this patent does not mention the Coriolis force, it does describe the following mechanism:

> A curved conduit, means for leading fluid to and from said conduit, means for imparting to said conduit angular movement about an axis transverse to an axis about which there exists angular momentum of the flowing fluid through the conduit, and means, including a device sensitive to gyroscopic couples of said conduit transverse to the axis of movement of the conduit, for indicating mass flow through the conduit.

Interestingly, this patent cites no previous patent references.

In addition to its use in meteorological circles, Coriolis' theory came to be applied to the development of flowmeters. The earliest patent that proposed setting up vibrations in flowing fluid was published by Ernest Fisher on behalf of the General Electric Company in 1917.

THE TERM "CORIOLIS FORCE" INTRODUCED

The earliest patent that includes the term "Coriolis Flowmeter" was granted in 1958 to Roby White. This patent was called CORIOLIS MASS FLOWMETER and it proposed subjecting the fluid to tangential acceleration in a whirling tube. This patent was filed on behalf of American Radiator & Standard Sanitary Corporation. This appears to be the earliest patent that mentions the "Coriolis force." The flowmeter is described as follows:

> The present invention relates to instruments for measuring the mass rate of flow of fluids and in particular to an improved flowmeter of the type in which mass flow rate is made responsive to Coriolis force… In instruments of the class described the fluid to be measured is subjected to tangential acceleration in a whirling tube, or impeller, the torque exerted on the tube in reaction to the Coriolis force of the accelerated fluid being measured as an indication of the mass flow rate.

ROBY WHITE – 1958

Patent Title: Coriolis Mass Flowmeter
Patent Number: US2,832,218
Filing Date: July 9, 1954
Issue Date: April 29, 1958

Roby White was the first to use the term "Coriolis Mass Flowmeter" in a patent. His design used actual rotation of fluid through a curved path. The Coriolis force was invoked directly by the angular motion of the fluid. Though the mechanism differs from modern vibrating-tube designs, this was the earliest application of Coriolis physics in flow measurement.

Historical Note: This was the first patent to use the Coriolis name in relation to flow measurement. It relied on actual rotation of the conduit.

YAO TZU LI – 1960

In May 1960, Yao Tzu Li patented an invention called "MASS FLOWMETER" that involves rotating the flow:

> The present invention operates by causing the fluid to be rotated as it flows radially outward from an axis. This produces a Coriolis acceleration in the fluid, and therefore a Coriolis force is applied by the fluid to the member through which the fluid flows. This force is measured, and the mass rate flow of the fluid is obtained.

Interestingly, Yao Tzu Li cites Fisher (1917), Kollsman (1952), and Pearson (1953) as references, but does not cite White (1958), who specifies a mechanism for rotating the fluid and also mentions the Coriolis force.

Neither this patent nor the Roby White patent were developed into products. Apparently the idea of producing Coriolis acceleration in a flowmeter by rotating fluid proved to be challenging from a practical point of view.

Patent Title: Mass Flowmeter
Patent Number: US2,934,951
Filing Date: September 9, 1952
Issue Date: May 3, 1960

Yao Tzu Li, a renowned MIT professor and aerospace engineer, developed an improved version of White's rotational design. His patent refined the mechanics of rotating conduits to more accurately measure mass flow using Coriolis principles. Though it still relied on mechanical rotation, Li's work demonstrated the feasibility and engineering maturity of Coriolis-based flow metering.

Li describes his meter as follows:

> The present invention operates by causing the fluid to be rotated as it flows radially outward from an axis. This produces a Coriolis acceleration in the fluid, and therefore a Coriolis force is applied by the fluid to the member through which the fluid flows.

This force is measured and the mass rate flow of the fluid is obtained. The present invention operates by causing the fluid to be rotated as it flows radially outward from an axis. This produces a Coriolis acceleration in the fluid, and therefore a Coriolis force is applied by the fluid to the member through which the fluid flows. This force is measured and the mass rate flow of the fluid is obtained.

Historical Note: Li helped formalize the mechanical and theoretical basis for early Coriolis meters using rotating tubes.

ANATOLE SIPIN – 1965 AND 1967

In November 1965, July 1967, and December 1967, Anatole J. Sipin patented three separate devices designed to measure mass flow. These designs by Sipin were designed to improve on earlier designs (described above) that involved rotating the flow or using an oscillating gyroscope. Sipin's objection to rotating gyroscopic designs was that they required rotating fluid seals, which introduced potential leakage and friction problems. His objection to oscillating gyroscopic designs was that they introduced bends and turns that made the meter difficult to clean. Meter size was also an issue. According to Sipin, "the diameter of the loop must be ten to twenty or more times as large as the diameter of the flow passage…(making) the apparatus large and cumbersome for the flow range." This problem of meter size is one that Coriolis manufacturers have never fully solved, except for meters two inches or less in diameter.

In 1965, Anatole Sipin was granted a patent called Mass Flowmeter Systems. He was critical of previous patents that called for rotating the fluid, saying these were too large to be practical. Instead, he proposed oscillating the fluid.

Patent Title: Mass Flow Measuring Device
Patent Number: US3,218,851
Filing Date: May 24, 1961
Issue Date: November 23, 1965

Anatole Sipin was the first to propose using oscillation instead of rotation in a mass flowmeter. His design involved vibrating a fluid-carrying tube and detecting the inertial effects of the moving fluid, which anticipates the key principles of modern Coriolis meters.

Historical Note: Sipin's concept of using vibrating tubes to detect mass flow by inertia laid the foundation for all subsequent Coriolis flowmeter designs.

1975 – JAMES E. SMITH

When Jim Smith came along and applied for his patent in 1975, he continued with the oscillation method introduced by Sipin in 1965. Even though this was a departure from the mechanical rotation of motion described by Coriolis, he kept the name "Coriolis," probably because of the 1958 Roby White patent called CORIOLIS MASS FLOWMETER. The following is a more detailed look at Jim Smith's patent.

- U.S. Patent 4,109,524 – Method and Apparatus for Mass Flow Rate Measurement.
- Filed: June 30, 1975; Granted: August 29, 1978.
- Continued with vibrating tube design (from Sipin; oscillation, not rotation).
- Retained the "Coriolis" name.

1977 – MICRO MOTION FOUNDED

- James E. Smith commercialized his design with Micro Motion.
- The term "Coriolis Flowmeter" became standard in industry.

Today, the name persists for historical and mathematical reasons, even though 'inertial mass meter' might be a more accurate descriptor.

JIM SMITH'S CORIOLIS PATENTS

In August 1972 James E. Smith patented a "Balanced Mass-Moment Balance Beam with Electrically Conductive Pivots." Beginning in June 1975, James (Jim) Smith began patenting a series of devices that became the basis for the flowmeters produced by Micro Motion. His patent filed in June 1975 was granted in August 1978. This patent and subsequent flowmeter patents explicitly evoke the Coriolis force. Smith's patents substituted a vibrating and oscillating tube for a rotating tube (following Sipin, 1965), which worked better than the earlier ones and is still used today in many different forms.

A little-known story about Jim Smith, as related by Dr. Mark Bell of Emerson, is that at the same time he developed the Coriolis meter, he invented another technology: an automatic nail gun. Dr. Bell relates the story as follows:

> He invented two technologies and put them both for sale, and said whichever one sells, he's going to develop the other one. He developed the first automatic nail gun. I got pictures of it. And he was like, for sure one of these chemical companies is going to buy the Coriolis meter patent, and I'm going to buy it, I'm going to develop the nail gun, and I'm going to become rich. Black and Decker bought the nail gun. It was a terrifying thing. You know how they throw baseballs with two spinning wheels? It was a two spinning wheel nail gun shooter… Black and Decker bought it, and it was almost immediately displaced by pneumatics, and then, electrodynamics, and some other things.

Jim Smith founded Micro Motion out of his garage in 1977. Micro Motion debuted its first Coriolis flowmeter in 1977 – an "A" meter for laboratory use. It was followed by the "B" meter in 1978. In 1981, Micro Motion introduced the first single-bent tube Coriolis meter, the "C" meter. Then in 1983, the company added the first dual-bent tube Coriolis meter, the "D" meter, which had a 2-inch diameter. In 1984, Emerson acquired Micro Motion. This proved to be one of the most successful acquisitions in flowmeter history.

BIBLIOGRAPHY

Gaspard Gustave de Coriolis, Sur les Equations du Mouvement Relatif des Systemes de Corps (Paris: Fondation Maison des Sciences de l'Homme, 2014)

Gaspard Gustave de Coriolis, *On the Equations of the Relative Motion of the Systems of Bodies*, Translated by Belinda Busteed Burum and Jesse Yoder (Wakefield, Massachusetts: Flow Research, Inc.: 2025)

6 Theory of Coriolis Flowmeters

In Chapter 5, we found that Gaspard Gustave de Coriolis wrote a very influential paper in 1835 that eventually became the basis for what is known today as the Coriolis force. While some commentators have stated that this article (which appears on www.worldflow.com) was about waterwheels (See Appendix), Coriolis does not mention waterwheels in his paper. Instead, he limits his discussion to rotating frames of reference. His main point is that in order to account for motion around a rotating frame of reference, it is necessary to introduce what he calls compound centrifugal forces. These compound centrifugal forces are what later became known as Coriolis forces.

It is also important to be aware that Coriolis did not discuss meteorology or what is now known as the "Coriolis effect" on weather systems in his article. So, how did the term "Coriolis" become attached to a certain type of flowmeter? The following timeline outlines the main steps that led to this result, as discussed in Chapter 5:

- This timeline illustrates how the term "Coriolis" became associated with mass flowmeter technology, even as the physical mechanisms shifted significantly.
- In 1958, Roby White patented a device that rotated the fluid flow and relied on the actual Coriolis force arising from angular motion. He explicitly called it a "Coriolis Mass Flowmeter."
- In the 1960s, other designs continued this trend of using mechanical rotation to invoke the Coriolis effect.
- In 1965, Anatole Sipin substituted oscillation for rotation and described achieving "Coriolis-like" forces.
- In 1975, Jim Smith applied for a new patent that build on Sipin's design. He described using oscillating tubes instead of rotating flow. Despite the lack of true rotation, he retained the term "Coriolis," likely due to continuity with earlier patents. This patent was granted in 1978.
- By 1977, with the founding of Micro Motion and commercialization of Smith's design, the term "Coriolis flowmeter" was fully established in industry – even though the underlying physics was better explained by fluid inertia than by true Coriolis forces.

Today, the name persists for historical and mathematical reasons, even though "inertial mass meter" might be a more accurate descriptor.

The patents granted in the 1950s and 1960s that described flowmeters that rotated the flow were not commercialized and were most likely not practical. This was the

DOI: 10.1201/9781003457855-6

criticism of Anatole Sipin, who had three different patents approved from 1965 to 1967. He makes the following comments in his 1965 patent:

> Other mass flowmeters that have obstruction-free flow passages are the rotating or oscillating gyroscopic meters of which several types are known. The gyroscopic meter has the disadvantage of requiring at least one and sometimes several loops with associated bends and turns between the flow inlet and outlet. For accurate flow measurement, moreover, the diameter of the loop must be ten to twenty or more times as large as the diameter of the flow passage (assuming circular cross-sections); and this requirement makes the apparatus large and cumbersome for the flow range… In addition, the rotating gyroscopic type of meter requires rotating fluid seals, introducing serious leakage and friction problems.

In place of earlier designs, Anatole Sipin in Patent 3,218,851 proposes using an oscillating design. This patent was filed in 1961 and approved in 1965. Sipin proposes the following:

> The mass flow measuring principle used in the embodiments of this invention is that where an oscillatory motion is applied by a member to a bounded stream of material and there is a changing difference between a transverse momentum, that is momentum orthogonal to the axis of flow, or the fluent material entering the oscillating member and that leaving the oscillating member, mechanical energy is removed from the oscillating member and added to the stream at a rate directly proportional to the mass rate of flow.

In discussing his invention, Sipin does not claim to be employing the Coriolis force, though he says that what he is doing is "equivalent to the Coriolis force" and "equivalent to the Coriolis acceleration." Furthermore, he called his patent "Mass Flowmeter Systems," not making an explicit reference in the title to Coriolis flowmeter.

When Jim Smith applied for his patent in 1975, he built on Sipin's idea of oscillation instead of rotation and proposed using the difference in phase shift between the incoming fluid and the outgoing fluid in a bent pipe to measure mass flow. Jim Smith in his patent, which was approved in 1978, called his invention "Method and Apparatus for Mass Flow Measurement." He cited three previous patents by Anatole Sipin as supporting documents. However, he drops Sipin's more cautious phrasing of "equivalent to the Coriolis force" and "equivalent to the Coriolis acceleration." In Smith's abstract, he appeals directly to the Coriolis force, saying that his device measures "the torque generated by the Coriolis force." At this point, the move from rotation to oscillation was complete, and Coriolis flowmeters were born as devices that oscillate the flow.

HOW CORIOLIS FLOWMETERS WORK

This section examines how Coriolis flowmeters work. It begins with the answers to some questions from a visit I made to Emerson Micro Motion in 2022.

QUESTIONS ABOUT HOW CORIOLIS FLOWMETERS WORK

As part of the research for this book, I visited Micro Motion and met with Dr. Mark Bell, one of the top Coriolis theorists at Micro Motion. He was kind enough to respond to several questions I (JY) had. Here are his (MB) responses:

JY: My main goal in visiting Micro Motion this time (I've been there before) is to better understand the operating principle of Coriolis flowmeters. I wrote about this in my book New-Technology Flowmeters (2023), but I still have several questions:

1. What is the relation between the Coriolis force and/or Coriolis effect and the operating principle of the Coriolis flowmeter?
   (MB) F = m*A, Newton's Law
   Coriolis effect is generally the particle acceleration (A), which is 2 ωxV, ω is angular velocity and V is linear velocity
   Coriolis force is Mass of fluid * Coriolis Acceleration
2. (JY) The Coriolis effect has to do with the behavior of objects on a rotating frame of reference. Is oscillation a form of rotation? I assume that if there is rotation, that it refers to the meter body, not to the fluid being measured
   (MB) ω is the rotation of the tubes while vibrating, the fluid mass is also rotating with the tubes
3. (JY) What is angular velocity and how is it calculated?
   (MB) Angular velocity is the rate of change of an angle, often measured in radians/second (degrees/second)
   It is calculated by measuring the rate of change of an angle

Dr. Bell goes on to explain that, in his view, the Coriolis force falls under Newton's law:

> The Coriolis force is then just the mass of the fluid times the Coriolis acceleration. That equals Newton's law. That's it. That's the only difference between the two, at least the way I think about it. You've got to make sure you're in the same coordinate systems and everything else. But yes, in general, forces are mass times acceleration. And this is how Coriolis meters work. And we apply that law in the non-rotating frame of reference. So, the fluid is rotating in the frame and we do mass times acceleration in a non-rotating frame.

The drive coil of a Coriolis flowmeter is usually located in the middle of the flowtube. The drive coil is an electromagnetic coil that causes the flowtube to vibrate. The flowtube oscillates in its natural sinusoidal motion as a result of the tube's vibration. Vibration and oscillation are similar but different concepts. Vibration is what the drive coil does to the flowtube. Oscillation describes the tube's natural sinusoidal motion in response to the force of the drive coil.

The pickoff sensors detect the motion of the flowtube as the fluid moves through it from inlet to outlet. Coriolis meters contain two pickoff sensors that are symmetrically located: one on the inlet side and one on the outlet side. These pickoff sensors contain a coil and a magnet. The magnet is attached to the flowtube.

HOW A CORIOLIS METER WORKS

As the magnet in a pickoff sensor moves through the coil, the moving magnet creates a changing magnetic field that interacts with the electrons in the wire, pushing them and creating a voltage (Figure 6.1). The amount of voltage induced depends on the rate of change of the magnetic field (how fast the magnet is moved), the strength of the magnetic field, and the number of turns in the coil. Because the magnet is attached to the flowtube, the motions of the magnet capture the degree to which the oscillating flowtube is deflected by the flowing fluid.

The induced voltage varies sinusoidally with the motion of the tubes, resulting in a sine wave output. The phase difference (time delay) between the sine waves produced by the inlet and outlet sensors is directly proportional to the mass flowrate of the fluid. A larger time delay indicates a higher mass flowrate. The Coriolis transmitter analyzes the phase difference, or time delay, of the sine waves produced by the input and output sensors, and uses this value to compute mass flow. Mass flow is directly proportional to this phase difference or time delay.

MASS FLOW MEASUREMENT

As the flowtubes vibrate, the relative movement between the coil and magnet in each pickoff induces a voltage. The voltage signals from the pickoffs are processed by the meter's electronics. When there is no flow, the pickoff signals are in phase.

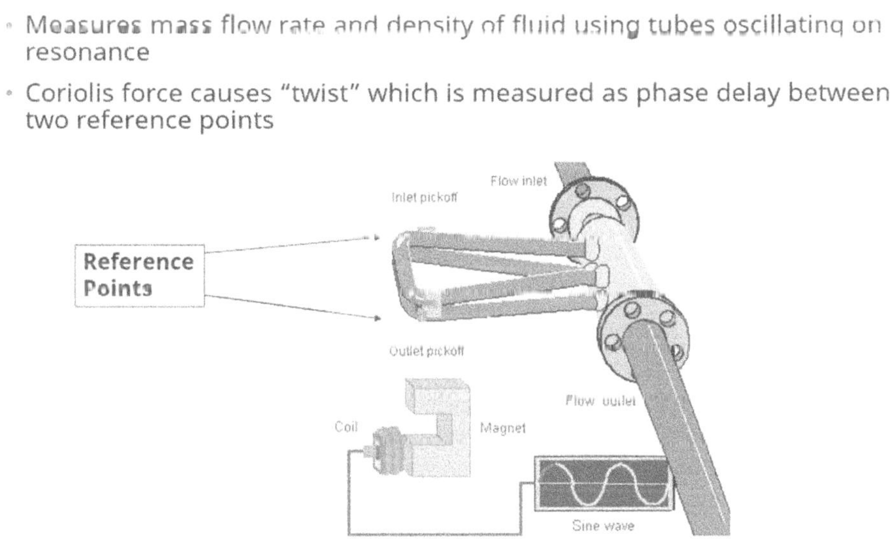

FIGURE 6.1 Coriolis flowmeter with inlet and outlet pickoffs.

Source: Courtesy of Emerson Micro Motion emerson.com/micromotion.

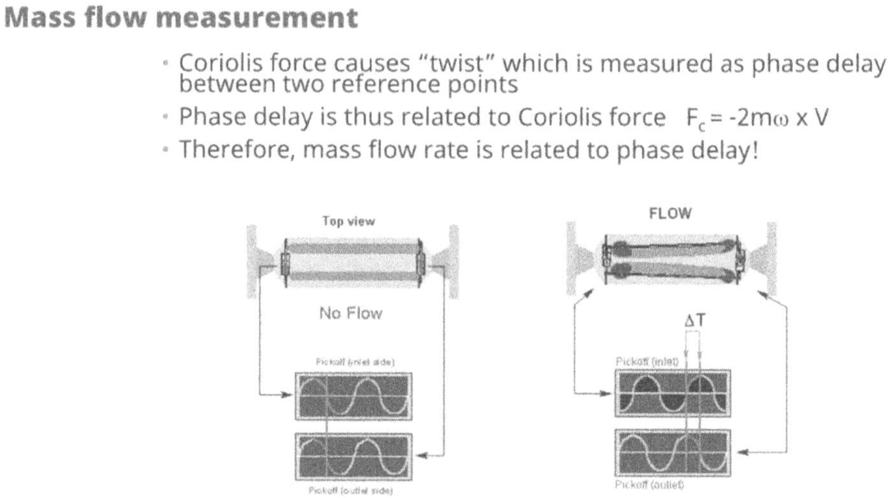

FIGURE 6.2 How the sine waves are generated by the pickoff sensors.

Source: Courtesy Emerson Micro Motion emerson.com/micromotion.

When fluid flows, oscillatory forces cause the tubes to twist, resulting in a phase shift between the pickoff signals. The magnitude of this phase shift is directly proportional to the mass flowrate of the fluid.

Due to the oscillatory motion of the flowtubes and the interaction of magnets and coils within the meter, the voltage is generated as a sine wave. In Figure 6.2 on the left side, the sine waves from the inlet and outlet pickoff sensors are in phase, since there is no flow causing deflection in the flowtube. On the right side, where there is flow, the inlet and outlet sine waves are out of phase, and the difference between the two is shown as a phase difference, or Δt. This difference in time between the sine waves from the inlet tube and the sine waves from the outlet tube is directly proportional to mass flowrate. The flowmeter transmitter uses this value to compute mass flow.

Figure 6.3 shows the inlet and outlet sensors at a different location from Figure 6.1, and also shows the action of the drive coil, which causes the flowtube to vibrate. The vibration of the flowtube results in the oscillation of the flowtube at its natural resonant frequency.

THE CORIOLIS EFFECT

As we learned in Chapter 5, Coriolis flowmeters are named after Gaspard-Gustave de Coriolis, a French mathematician and engineer. In 1835, he published a paper in which he described the behavior of objects in a rotating frame of reference. While this is sometimes called the Coriolis force, it is more accurately called the Coriolis effect, since it is not the result of a force acting directly on the object, but rather the perceived motion of a body moving in a straight line over a rotating body or frame of reference.

One common example given to illustrate the Coriolis effect is that of a ball propelled through the air a long distance in a straight line from the North Pole toward

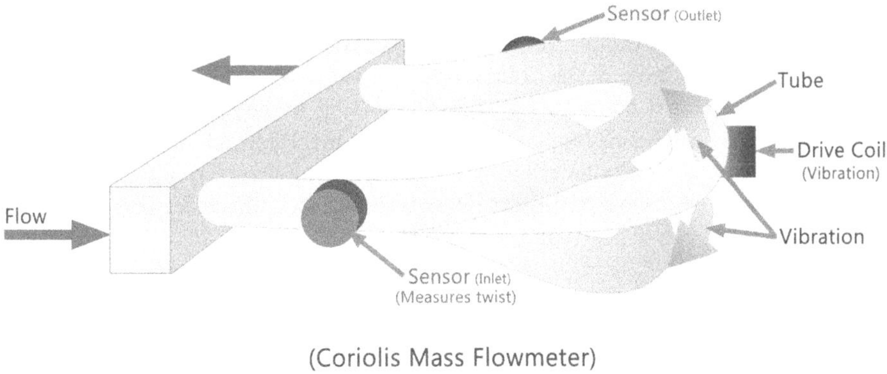

FIGURE 6.3 A Coriolis flowmeter showing the inlet and outlet sensors, and the action of the drive coil.

a target on the equator. When the ball arrives at the equator, it will not land at its target, because the Earth will have rotated underneath the moving ball so that it will land a distance away from the perceived target on the equator. The ball will appear to have curved from the perspective of the person standing where the ball is "thrown."

A similar example is a person throwing a ball from the center of a rotating merry-go-round. If she throws the ball at a horse on the outside edge of the merry-go-round, she will not hit the horse she is aiming at but perhaps the one after it. By the time the ball reaches the edge of the merry-go-round, the horse she aimed at will have moved to another location.

In both these cases, there is no force acting to push the ball in a curved direction. Instead, its motion will appear to be curved from the perspective of the ball-thrower because the frame of reference is moving underneath the ball. For this reason, the Coriolis "force" is more appropriately called the Coriolis effect. It refers to the apparent effect on the motion of an object passing over a rotating frame of reference when viewed from the point of origin of that moving object.

HOW THE CORIOLIS EFFECT OCCURS ON THE EARTH

The Coriolis effect occurs in rotating systems, such as the Earth, which rotates around its axis. In a rotating frame, an object moving in a straight line will appear to follow a curved path. This is due to the rotation of the reference frame itself. On Earth, this means that as air or water moves over the surface, it is deflected to the right in the Northern Hemisphere and to the left in the Southern Hemisphere. The Earth rotates from west to east, causing different points on its surface to move at different velocities. Points on the equator move faster than points near the poles. Objects moving across the rotating Earth retain their initial velocity relative to space. As the Earth rotates underneath them, their paths appear to curve. The Coriolis effect is stronger at the poles and weaker at the equator. This is because the rotational velocity of the Earth is greatest at the equator and zero at the poles.

The Coriolis effect influences wind patterns and the rotation of large-scale weather systems. For example, hurricanes in the Northern Hemisphere rotate counterclockwise, while those in the Southern Hemisphere rotate clockwise. Ocean currents are also affected by the Coriolis effect, leading to the development of gyres, which are large circular current systems in the oceans. Pilots and missile engineers must account for the Coriolis effect when plotting long-distance flights or missile trajectories, as it can cause significant deflection over large distances (Figure 6.4).

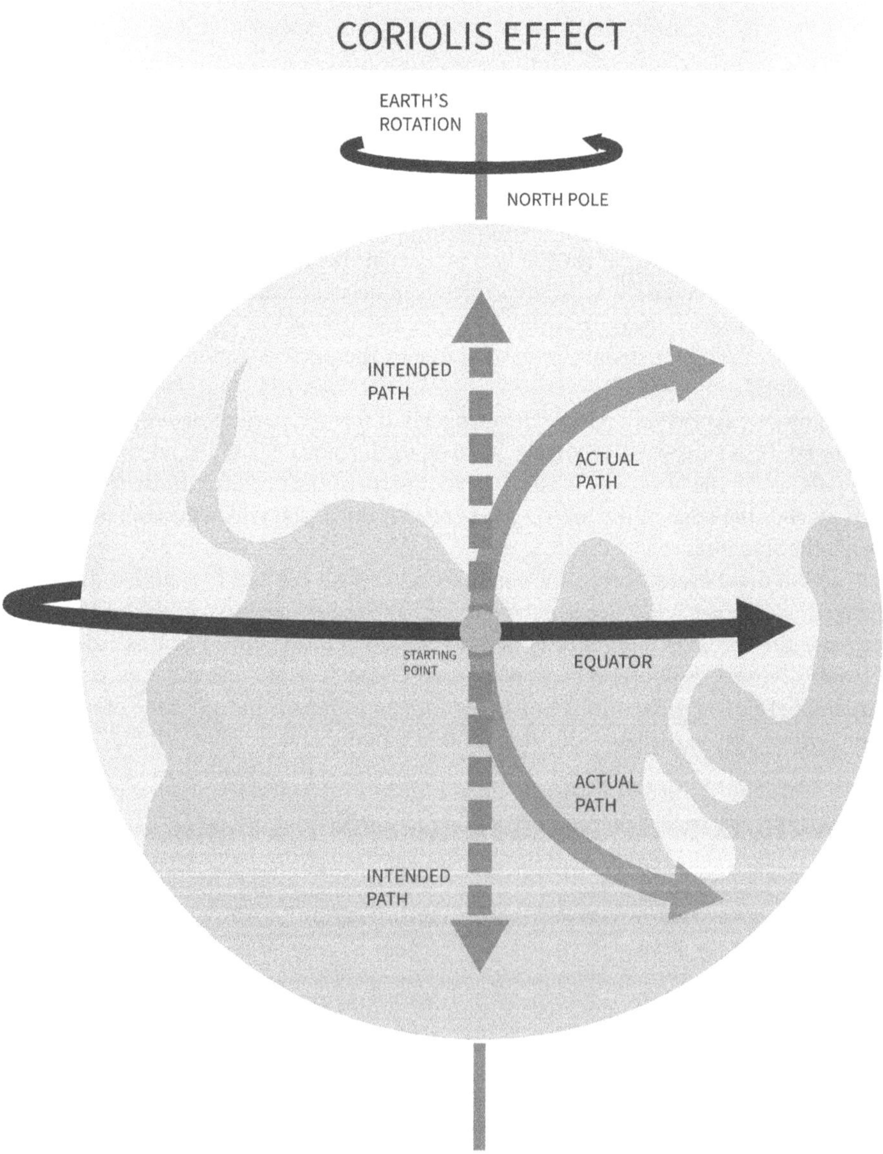

FIGURE 6.4 The Coriolis effect as it applies to the Earth.

THE CORIOLIS EFFECT AS APPLIED TO CORIOLIS FLOWMETERS

In the examples of the Coriolis effect commonly given, this occurs in rotating systems that are not attached to the frame of reference. In the case of planes and large-scale weather systems, they are traveling over the rotating Earth. If we hypothetically throw a ball from the North Pole to a target on the equator, the ball flies over the rotating Earth and lands somewhere west of its target. While the ball is in flight, the Earth is rotating from west to east. The merry-go-round example is similar. As a ball is thrown through the air from one side of a merry-go-round to the other, it will land somewhere beyond where the horse aimed at is. This is because the merry-go-round is rotating underneath the flying ball.

In all these examples, the flying or thrown object is traveling above the rotating frame of reference; it is not part of the frame of reference. If the thrown object were part of the rotating frame, we would not experience the Coriolis effect because the thrown object would be moving with the rotating frame. For example, if we rolled a ball across the merry-go-round instead of throwing it through the air, chances are good that it would hit its target, since it would be moving along with the horse the ball is aimed at. Likewise, we don't experience the Coriolis effect with cars traveling from south to north, since they simply rotate with the Earth along with their target cities. Someone traveling from Boston, Massachusetts, to Boothbay Harbor, Maine, doesn't have to adjust for the Coriolis effect because the car, the road, and the town in Maine are all rotating with the Earth.

If this is the Coriolis effect, then it is difficult to see how it applies to a Coriolis flowmeter. The fluid traveling through an oscillating tube is not traveling over a rotating frame of reference; it is actually a part of the frame of reference it is traveling through. The fluid has inertia, and as the fluid with inertia encounters oscillation, it is accelerated back and forth with forces perpendicular to its direction of travel. However, the fluid becomes part of the oscillating frame of reference, since the tube oscillations create oscillations in the fluid. This example is not parallel to the usual examples of the Coriolis effect, where an object travels above a frame of reference and does not travel with that frame of reference.

OSCILLATION VS. ROTATION: ARE THEY THE SAME OR DIFFERENT?

Oscillation and rotation are distinct types of motion. Rotation involves movement in a circle around a central point or axis, while oscillation is a back-and-forth or to-and-fro movement around a central point or equilibrium position. Essentially, rotation is a circular motion, while oscillation is a repetitive, linear or curved back-and-forth motion.

The differences can be explained more clearly as follows (Figure 6.5).

Rotation

- Involves a circular path around a fixed point or axis.

FIGURE 6.5 A Ferris wheel.

- Can be non-uniform (changing speed) or uniform (constant speed).
- Examples include the Earth's rotation on its axis, a spinning top, a merry-go-round, or a Ferris wheel.

Oscillation

- Involves repetitive movement back and forth or up and down.
- Typically occurs around a central point of equilibrium.
- Examples include a pendulum swinging, a vibrating string, or a spring bouncing (Figure 6.6).

FIGURE 6.6 An oscillating pendulum.

Main Differences:

- **Direction:** Rotation maintains a consistent direction of circular movement, while oscillation involves reversing direction at the extremes of its movement.
- **Path:** Rotation follows a circular path, while oscillation follows a back and forth path.
- **Examples:** Rotation can be seen in the movement of a wheel, while oscillation is seen in the movement of a spring, a metronome, or the pendulum of a clock.

Some people say that oscillation and rotation are the same, or that oscillation is a form of rotation. From this point of view, going part of the way around a circle is rotating, and going the other way is rotating in a different direction. When the Coriolis tubes are moving apart, this tube is rotating that way and this tube is rotating that way. While this is one way to look at the situation, it involves stretching the meaning of the term 'rotation' to fit a theory. It's not completely incorrect, but "oscillation" is a more accurate description of the situation than "rotation." And we know that Coriolis flowmeters are all about accuracy – to within 0.05%!

WHAT HAPPENS TO THE FLUID WITHIN THE TUBES

In a Coriolis flowmeter, the fluid flows through tubes that are vibrating perpendicular to the direction of fluid flow. These vibrations are typically sinusoidal and are driven by a drive coil. As the fluid moves through the oscillating tubes, each fluid particle is accelerated due to the vibration.

The acceleration experienced by a fluid particle due to the vibration of the tubes in a Coriolis flowmeter is called centripetal acceleration. This type of acceleration occurs because the vibrating motion of the tubes forces the fluid particles to follow a curved path creating a centripetal force that acts toward the center of the curvature of the motion. The motion is essentially a repeated back-and-forth curvature of the path. The fluid particles must constantly change direction to follow the vibrating tube, and these changes in direction require centripetal acceleration.

As the fluid flows through the oscillating tube, a secondary vibration occurs in the tube. This secondary vibration is a twisting motion that occurs because the particles are trying to go in a straight line. As described above, inlet and outlet sensors detect this induced twisting motion as a phase shift between the inlet and outlet sensors. The magnitude of this phase shift is proportional to mass flowrate.

FLUID INERTIA PLUS OSCILLATING FORCES YIELD A MASS INERTIAL EFFECT

While the Coriolis force is defined in the context of rotating frames, a similar inertial effect arises in oscillating systems like those in Coriolis flowmeters. This effect is due to the changing velocities and directions within the oscillating frame of the tubes. The twisting motion observed in Coriolis flowmeters is a result of those inertial effects, akin to the Coriolis effect, even though the tubes are not rotating continuously.

The traditional Coriolis force applies to objects in a rotating frame of reference, causing apparent deflection due to relative motion. Oscillating tubes in flowmeters generate inertial effects similar to the Coriolis force when fluid flows through them, resulting in a measurable twisting motion. The induced twist or phase shift in the tube caused by these inertial effects allows the flowmeter to determine the mass flowrate. Thus, the Coriolis flowmeter utilizes the principles of inertial forces arising from oscillatory motion to achieve a similar effect to the Coriolis force, enabling accurate flow measurement.

AN ALTERNATIVE UNDERSTANDING OF THE CORIOLIS FLOWMETER

The vibrating tubes create centripetal acceleration, forcing fluid particles to follow the oscillatory motion. The fluid particles experience inertial forces due to the combination of their linear flow through the tubes and the oscillatory motion of the tubes.

TABLE 6.1
Two Alternative Views about How Coriolis Flowmeters Work

Traditional View (Coriolis Explanation)	Alternative View Inertial Mass Explanation
In the traditional view, Coriolis flowmeters are explained using the Coriolis effect in a rotating frame of reference. As fluid flows through the oscillating tubes, it experiences Coriolis acceleration. This causes the tubes to twist. The resulting phase shift between the inlet and outlet sensors is proportional to mass flowrate.	In the alternative view, the same tube twist is attributed to the inertial response of the flowing mass to the oscillating motion of the tubes. No rotating frame is required. Newton's second law (F = ma) is applied directly. The phase shift between the inlet and outlet sensors is still proportional to mass flowrate, but it is seen as a direct result of inertia – hence the name inertial mass meter

Instead of invoking the Coriolis effect in a rotating frame, I believe these meters work on a simpler, more direct principle: the inertial response of flowing mass to oscillating acceleration.

(Yoder, 2011)

These inertial forces result in a secondary twisting motion in the vibrating tubes. This twist is a measure of the mass flowrate of the fluid.

Inlet and outlet sensors detect the phase shift caused by the induced twisting motion. The phase difference detected by the inlet and outlet sensors is directly proportional to mass flowrate.

In the context of a Coriolis flowmeter, the observed twisting motion is due to inertial effects arising from the interaction of fluid flow and tube oscillation. While these effects are similar to the Coriolis force, they do not strictly involve the Coriolis force as it applies to rotating frames (Table 6.1).

BIBLIOGRAPHY

Jesse Yoder, *Unquestionable Accuracy, but Are Coriolis Meters Aptly Named?* (Birmingham, Alabama: Flow Control, Grand View Media, November, 2011)

7 Development of Coriolis Flowmeter Products

THE FIRST EUROPEAN CORIOLIS FLOWMETER

Heinrichs Messtechnik GmbH says on its website that it is the first European company to offer Coriolis flowmeters. Heinrich introduced its first Coriolis meter in 1986, which was the same year that Rheonik debuted its Omega-shaped Coriolis meter. That was also the year Endress+Hauser released its first Coriolis meter. This meter was a dual-tube, straight-tube meter, which evolved into the Proline Promass F. As a result, it is open to debate about which company was the first European company to introduce a Coriolis flowmeter.

EMERSON AND EXAC CORPORATION

Micro Motion Coriolis meters have maintained their leadership position in Coriolis flowmeters in part by vigorously defending patents. In 1983 and 1984, Exac Corporation began designing and marketing Coriolis meters. In 1987, Emerson sued Exac for patent infringement. The initial finding of the court was that Exac's patents did not infringe on Micro Motion's patents. However, in December 1987, the court granted Micro Motion's request for a new trial. The second trial began in March 1990. Micro Motion based its case on lost profits due to lost sales. The calculations of lost sales and profits due to the court's finding of Exac's patent infringement is too complex to review here, but public records exist for this. In the end, Exac Corporation was ordered to pay Micro Motion $26.2 million. More specifically, in 1991, the court found infringement under the doctrine of equivalents and awarded Micro Motion $20,821,525 in damages plus $5,409,481 in prejudgment interest, totaling $26,231,006. The decision was called the decision in *Micro Motion, Inc. v. Exac Corp.*, 761 F. Supp. 1420.

Exac Corporation continued to exist as a subsidiary of Fisher Controls International (a Monsanto subsidiary) during litigation. However, there is little evidence that Exac continued to operate long-term afterward – public filings and case reporting indicate the firm effectively faded from the marketplace in the years following the ruling. No prominent later filings or product releases are documented post-early 1990s.

Available court records confirm the judgment amount, but they do not provide clear public documentation about whether Exac actually paid in full or at all, or whether it negotiated settlement, declared bankruptcy, or had insurers absorb the cost. There do not appear to be credible sources verifying full payment or enforced collection.

More importantly, Micro Motion has followed up with innovative designs even when other companies were first to the market. For example, KROHNE brought out

DOI: 10.1201/9781003457855-7

the first commercially successful single straight tube meter in 1994. Micro Motion responded with its own straight tube meter in 1999. Technological development and product improvement have consistently been hallmarks of Emerson Micro Motion's approach to Coriolis meters. The company's closest competitor is Endress+Hauser.

KROHNE'S SINGLE STRAIGHT TUBE CORIOLIS METER

Schlumberger was next to introduce a straight-tube flowmeter in 1993, but withdrew this product after several months. KROHNE introduced the first commercially successful single straight-tube Coriolis meter in 1994. Since that time, this type has become increasingly popular with Coriolis users. Straight-tube meters address the problem of pressure drop because the fluid doesn't have to travel around a bend, making them better able to handle high-viscosity fluids. One of the moving forces behind Krohne's meter was Dr. Yousif Hussain, who holds several patents on Coriolis meters.

In 1984, Rheonik's founder, Karl Küppers, began developing what was to become the company's technology base and patent portfolio. The design was the patented Omega-shaped Coriolis flowmeter. Rheonik was founded in 1986, the same year it introduced its Coriolis meter. In 1993 due to growth Rheonik moved to its current facility in Odelzhausen, Germany. Rheonik brought out the world's first 12-inch Coriolis meter in 2000. At that time, it was the largest Coriolis meter made. In February 2008, GE Sensing and Inspection Technologies acquired Rheonik. However, Rheonik purchased back the assets and business from the GE unit in 2015 (Figure 7.1).

In 1992. Micro Motion introduced its Elite series of Coriolis meters. This line was designed for flow and density measurements of liquids, gases, and multiphase flow.

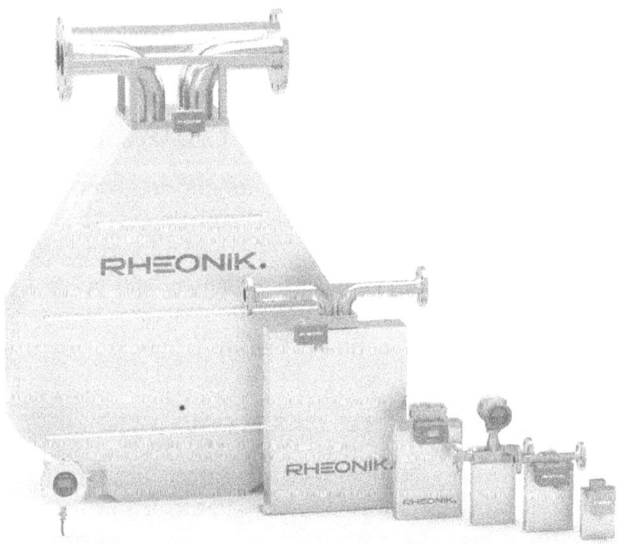

FIGURE 7.1 Range of Rheonik's Coriolis meters.

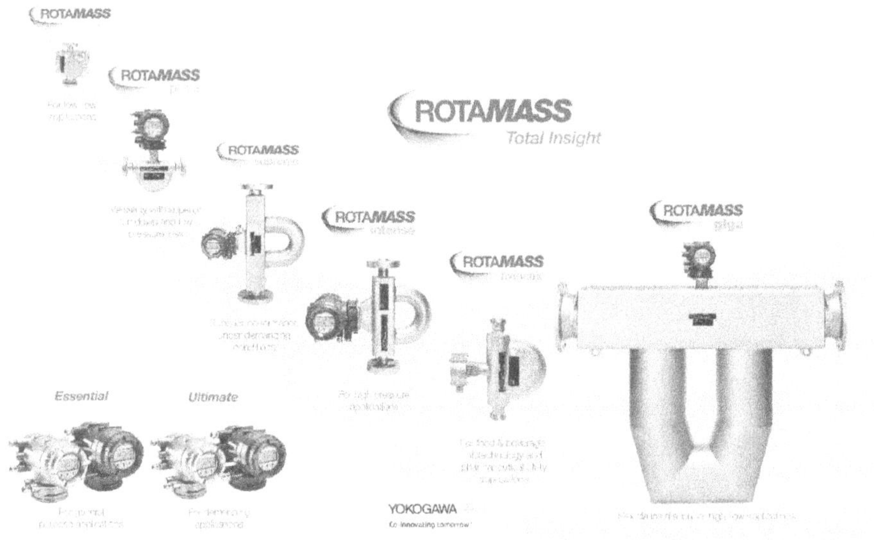

FIGURE 7.2 Range of Yokogawa's ROTAMASS Coriolis meters.

Emerson still carries its Elite series meters. In 1995, the company brought out its F-Series line, is compact and drainable, and has a robust design (Figure 7.2).

During this same period, in 1993, the company Rota in Germany brought out the ROTAMASS, which was developed in Wehr, Germany. This dual-tube flowmeter addressed the effects of vibration or pipeline stress with a heavy wall. It also provided increased reliability and output stability. In 1995, Rota became a subsidiary of Yokogawa Europe.

MICRO MOTION RESPONDS TO OTHER COMPANIES' PRODUCTS

Micro Motion has followed up with innovative designs even when other companies were first to the market. For example, KROHNE brought out the first commercially successful single straight tube meter in 1994. Micro Motion responded with its own straight tube meter in 1999. Technological development and product improvement have consistently been hallmarks of Emerson Micro Motion's approach to Coriolis meters. The company's closest competitor is Endress+Hauser.

STRAIGHT-TUBE, RADIAL-MODE GAS FLOWMETER

Seven ex-employees of Micro Motion founded Direct Measurement Corporation (DMC) in 1991. The company's focus was on Coriolis flowmeters for the oil and gas industry. In 1996, FMC Technologies, now Guidant Measurement, acquired DMC. DMC's main product was the straight-tube, radial-mode Coriolis meter for measuring gas flow.

By 2000, FMC had three different Coriolis meters: S-Mass, Apollo A400, and the radial-mode flowmeter. Both the S-Mass and the Apollo A400 were Smith Meter brands and designed for liquid applications. The S-Mass meter was named for the S-shaped design of its flowtube. Available models included the S25, S50, S100, S200, and S300. The S-Mass was designed for a variety of applications including custody transfer, blending, leak detection, batch control, online density measurement, and petroleum production.

The Apollo A400 was designed for use in petroleum applications. It measured mass flow at rates up to 20,000 lbs/min. The A400 was available with the HART communication protocol. Applications include loading rack terminals or bulk deliveries, transportation of crude oil and refined products, and LACT systems.

FMC had two models of the radial-mode flowmeter for measuring gas flow: the R200, a 2-inch meter and the R400. The R400 was a larger meter with greater throughput capabilities. FMC was still selling the meter in 2003 but discontinued it a few years later.

It is challenging to measure gas flow with a straight-tube meter. But what are the benefits of a straight-tube meter that measures gas flow? One of the benefits of a straight-tube meter is that liquid doesn't build up at the curves. This does not apply to gas meters since gas does not collect at the curves in a bent tube meter. Also, bent-tube meters cause pressure drop. But gas causes minimal pressure drop, even in bent-tube meters. Pressure drop is a more significant consideration for liquids. The benefits of using a straight-tube meter to measure gas flow are limited, although they may be better suited to hygienic and sanitary applications than bent-tube meters.

WHY THE DUAL-TUBE CORIOLIS FLOWMETER WAS INVENTED

Based on research into the patents underlying the flowmeters introduced in the early 2000s, there are patents by Don Cage assigned to Micro Motion, DMC, and FMC Technologies in the period from 1995 to 2004. He is the inventor of the radial straight-tube Coriolis meter from DMC, which was bought by FMC Technologies. He continued to be active in Coriolis design for other companies after this period. Cage has been highly influential in the design and development of many Coriolis flowmeters over the past 30 years, including the TRICOR manufactured by Kem Kueppers.

One of the most interesting quotes from one of Cage's 1995 patents explains the invention of the dual-tube meter. The patent is called "Coriolis Mass Rate Flow Meter."

> It is well known that a vibrating flow conduit carrying mass flow causes Coriolis forces which deflect the flow conduit away from its normal vibration path proportionally related to mass flow rate…
>
> This effect was first made commercially successful by Micro Motion Inc. of Boulder Colo. Early designs employed a single vibrating U-shaped flow conduit which was cantilever mounted from a base. With nothing to counterbalance the vibration of the flow conduit, the design was sensitive to mounting conditions and so was redesigned to employ another mounted vibrating arrangement which functioned as a counterbalance for the flow conduit. Problems occurred, however, since changes in the specific gravity of the process-fluid were not matched by changes in the counterbalance, an unbalanced condition could result causing errors. Significant improvement was made later by replacing the counter-balance arrangement with another U-shaped flow conduit

identical to the first and splitting the flow into parallel paths, flowing through both conduits simultaneously. This parallel path, Coriolis mass flow rate meter solves this balance problem and has thus become the premier method of mass flow measurement in industry today.

There has been intense development of Coriolis technology since it was first introduced in 1977. Today, dual bent tube technology is the dominant configuration for Coriolis meters. Even though some companies offer single bent tube meters, as well as single straight and dual straight tube Coriolis meters, the bent tube configuration has won out in the large majority of cases, mainly because it is more stable and accurate than the alternatives. Today more than 40 companies manufacture Coriolis flowmeters.

MICRO MOTION, ENDRESS+HAUSER, AND KROHNE RESPOND TO RHEONIK

Once Rheonik introduced its 12-inch meter, the potential of the large line size Coriolis meter market did not go unnoticed by other companies. Emerson Process Management, Micro Motion division, introduced its first Coriolis flowmeter for line sizes above 6" in December 2007. Endress+Hauser and KROHNE have also introduced their own Coriolis meters in the large line size (> 6") range. More recently, Shanghai Yinuo, a Chinese manufacturer, has brought out large line size meters. Many of these are designed for custody transfer of petroleum liquids and/or natural gas. While the price tag for these meters can be quite high (e.g., $75,000), they can measure both types of fluids with very high accuracy. For this reason, they are prized by end-users. Even with the larger line sizes, Coriolis meters find it difficult to compete with ultrasonic meters for custody transfer of natural gas, which can measure flow in pipes with 16 and 20-inch diameters and beyond.

What is the upper limit of Coriolis flowmeters in terms of size? The limit seems to be both practical and theoretical. On the practical side, large line size Coriolis meters are both very big and very heavy. This adds to the expense of shipping them, adds to installation costs, and limits where they can be securely installed. On the theoretical side, it is more difficult to make a 14" Coriolis meter vibrate at the desired level than a 2" Coriolis meter. This doesn't mean that 14" or 16" is an absolute ceiling in terms of size for Coriolis meters; it only means that no company has yet produced a larger one. Interestingly, both vortex and positive displacement meters also currently top out at 16".

EMERSON MAINTAINS ITS MARKET LEADERSHIP

In July 2020, Emerson announced it was investing more than $100 million in Boulder, Colorado – the home of Micro Motion products – to significantly expand its manufacturing space and launch a new innovation center focused on research, new product development and industry training for its advanced flow measurement products. The facility was opened in January 2023 to host customer site visits. Micro Motion products are manufactured in Boulder, Colorado, as well as in the Netherlands, Mexico, China, and Romania (Figure 7.3).

FIGURE 7.3 Emerson's ELITE Series Coriolis meters.

Micro Motion, like Rosemount, is now an Emerson brand in the Measurement & Analytical division. Offering Coriolis flowmeters and complementary density and viscosity meters, Micro Motion has been the worldwide market leader in supplying Coriolis flowmeters for decades, although other vendors have encroached on this position.

The Micro Motion brand continues to dominate the worldwide Coriolis flowmeter market. Micro Motion has maintained a large market share in the Coriolis flowmeter business by continuing to offer product innovations, and by bringing out new Coriolis products as sizable opportunities arise (Figure 7.4).

The Micro Motion LNG Dispenser System, for instance, takes advantage of the growing demand for this capability, particularly now in Western Europe, where the LNG infrastructure is being built out country by country. Similarly, the compressed natural gas dispensing meter targets CNG users.

The company also benefits from Emerson's broad line of flowmeters, temperature and pressure transmitters, and valves. This wide range of hardware capabilities, combined with networking that ties them together in advanced process management control systems, makes the company competitive for large projects requiring

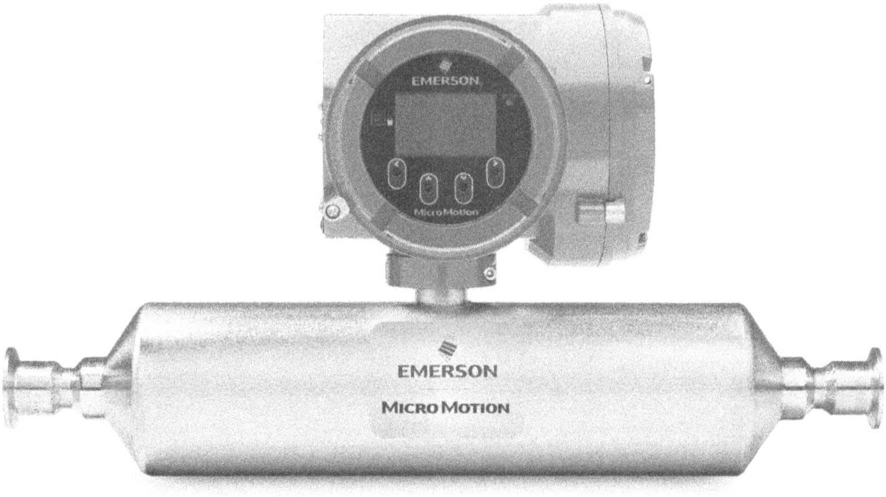

FIGURE 7.4 Emerson Micro Motion's T Series straight Tube Coriolis meter.

multiple instrumentation and control solutions. In addition, Micro Motion benefits from Emerson's worldwide sales and distribution network.

ENDRESS+HAUSER

Endress+Hauser is present in 125 countries, with manufacturing plants in Brazil, China, France, Germany, India, Italy, Japan, Switzerland, the United Kingdom, and the United States. One-third of the group's employees work in the tri-border area of Switzerland, Germany, and France (Figure 7.5).

Endress+Hauser's Coriolis flowmeters under the Proline product family name (introduced in 1993) combine various sensors with state-of-the-art transmitter technologies, and permit the accurate measurement of virtually all fluids, including cleaning agents, solvents, fuels, crude oil, refined oil products, vegetable oils, animal fats, latex, silicon oils, alcohol, fruit solutions, toothpaste, vinegar, ketchup, mayonnaise, gases, and liquefied gases.

The company's Coriolis flowmeters simultaneously measure mass flow, volumetric flow, density, temperature, and viscosity. There are no requirements for upstream or downstream runs for measuring either liquids or gases, and accuracy can be as high as ±0.05% of flowrate, with either four-wire or loop-powered configurations.

Endress+Hauser has made a practice of offering a wide range of Coriolis flowmeter designs for specific applications. For example, the Promass P 100 is useful in the life sciences industry due to its ultracompact transmitter and certification for hygienic applications. A single-tube design, made for line sizes ranging from $^1/_{24}$" to $^1/_8$" in diameter, it is suitable for installation with modular or skid-mounted process facilities, and is fully self-drainable even in horizontal installations. All wetted parts are stainless steel with electropolished surface finishes, and the engineering ensures full operational compliance through integrated self-verification. The Proline 100,

FIGURE 7.5 Endress+Hauser Coriolis flowmeters in use.

an ultra-compact transmitter version for hygienic applications, delivers full performance on a small footprint, enabling skid builders, systems integrators, and others to use limited space more efficiently.

In May 2011, the company unveiled the world's first commercially available 14-inch diameter, 4-tube high accuracy Coriolis flowmeter: the Promass 83X/84X for pipeline custody transfer of both gas and oil and other high flowrate applications. Today, the Promass X 300 model is a four-tube device in line sizes from 12" to 16" in diameter. It is used in onshore and offshore applications in the oil & gas industry, typically in pipelines and transfers to and from tanker ships, trucks, and railroad cars. The patented four-tube design provides accuracy up to 0.05% of flowrate.

The Promass Q 300 Series is specially designed for use in custody transfer, mass and volume flow, temperature, and density measurement in challenging conditions.

It also provides optimized performance for liquids with entrained gas using multi-frequency technology. Optional wireless and WLAN communication methods are available.

The company's Promass F 200 Series is designed for industrial applications for liquids and gases in the chemical and oil & gas sectors. This series is equipped with high-pressure sensor housings, purge connections, and rupture disks, with dedicated wetted parts for chemical and other process applications. It accommodates loop-powered capability and can be installed in hazardous areas due to its intrinsically safe design and approvals.

Promass F 300 has a long-standing reputation as a highly accurate sensor. Immune to fluctuating and harsh environments, it is suited for the broadest range of applications. With its compact transmitter, the Promass F 300 offers high flexibility in terms of operation and system integration: access from one side, remote display, and improved connectivity options. Heartbeat Technology ensures compliance and process safety at all times.

The Proline series is designed with future applications in mind, including accordance with SIL requirements for maximum safety, quality, and operational availability. Unique diagnostic functions and a sophisticated data storage concept also help to ensure these standards. The Proline is prepared for digital networking within process automation applications and can transmit data on either wired or wireless networks.

At the other end of the spectrum are the ultra-compact Cubemass and Promass A lines. The Cubemass C Series is used primarily in the automotive industry on test rigs measuring fuel consumption. It measures gas or liquid mass flow through line sizes below ⅛ inch even under high-pressure conditions.

The CNGmass series is used for compressed natural gas refueling applications, the LPGmass series for liquefied petroleum gas applications, and the LNGmass series for cryogenic gases applications.

Endress+Hauser claims that there are more than 800,000 of its Coriolis flowmeters installed worldwide. Endress+Hauser's Coriolis flowmeters under the Proline product family name introduced in 1993 combine various sensors with state-of-the-art transmitter technologies, and measure virtually all fluids, including cleaning agents, solvents, fuels, crude oil, refined oil products, vegetable oils, animal fats, latex, silicon oils, alcohol, fruit solutions, toothpaste, vinegar, ketchup, mayonnaise, gases, and liquefied gases. The company's Coriolis flowmeters simultaneously measure mass flow, volumetric flow, density, temperature, and viscosity. The high-capacity four-tube Proline Promass X 500 designed for the oil & gas industry is available in diameters up to 16 inches.

Endress+Hauser has made a practice of offering a wide range of Coriolis flowmeter designs for specific applications. Its Proline Promass U 500 meter, for example, is the first single-use flowmeter for biotech applications, offering premium performance, cGMP compliance, and up to 4 I/Os.

KROHNE

In 1979, KROHNE established operations in the U.S. The U.S. headquarters moved to a new facility in Beverly, Massachusetts in September 2018 next to a new manufacturing center for flow and level products. Production and distribution operations also followed to the new location. This $25 million facility features 90,000 square feet of office space and is staffed by more than 100 employees. Products manufactured there include Coriolis, magnetic, and variable area flowmeters, as well as radar and guided radar level transmitters.

In January 2020, Michael Rademacher-Dubbick, who successfully led the KROHNE Group for 25 years, became chairman of the advisory board. The KROHNE Group is now led by Dr Attila Michael Bilgic (CEO), Dr Marco Rudelli (CSO), Gaëlle Hotellier (COO), and Ingo Wald (CFO). The generational change ensures that the KROHNE Group will continue to be successfully positioned to meet the challenges of the market in the future. The year 2021 marked the 100th anniversary of KROHNE, and the aim of the new management team is to strengthen the foundation to set the course for the next 100 years.

KROHNE manufactures Coriolis flowmeters for all process and custody transfer applications, including mass, volume flow, density, and concentration measurement of liquids and gases. The company offers these flowmeters in various designs ranging from twin bent to single straight tube sensor types. In 1994, KROHNE was among the first companies to offer a commercially available straight tube Coriolis flowmeter. In July 2020, the company announced that its entire line of Coriolis OPTIMASS x400 flowmeters is available with Bluetooth Low Energy (BLE) capability, making these the only certified mass flowmeters on the market today so equipped.

KROHNE manufactures Coriolis meters for all process and custody transfer (CT) applications that measure mass, volume flow and temperature of liquids and gases, and density and concentration measurement of liquids. It offers secure wireless access via Bluetooth®, even in safety-related applications, and Entrained Gas Management (EGM™) that maintains operation over a wide range of gas volume fractions – even up to 100% – and complex flow conditions (Figure 7.6).

The OPTIMASS family covers the widest range of Coriolis products that KROHNE offers. The twin bent tube OPTIMASS 6400 includes a comprehensive diagnostics package that monitors the process and process conditions as well as the flowmeter. The high-capacity OPTIMASS 2400 is available in twin or four straight-tube designs, with flange connections up to 16 inches. It is suited for bunkering operations and custody transfer or fiscal metering in the oil and gas industry.

The 6400 is also able to provide seven measurement functions, specifically: mass flowrate, mass counter, density, temperature, volume flowrate, volume counter, and concentration of gases. A comprehensive diagnostics package monitors not only the flowmeter, but also the process and process conditions.

The Straight tube KROHNE series includes the OPTIMASS 1010, 1400, 2400, 7010, and 7400. Each series is strengthened for specific applications. For example, the 2400 is designed for bulk measurement of gases and liquids, maintaining effective operation over a wide range of gas fractions, and complex flow conditions. The 7400 has high accuracy, at ±0.1%, for use with viscous or aggressive fluids, is certified for use in hygienic applications, and is only available in line sizes of ½" to 4" in diameter – typical for sanitary fluid measurement environments.

The MFC400 mass flow signal converter, an important element of the company's Coriolis product line, features stability, an advanced density measurement,

FIGURE 7.6 KROHNE's OPTIMASS Coriolis meters in use.

push-button input, and an optical display. And, in the event of a fault within the electronics system, the complete electronics cartridge can be replaced in situ, and the standard redundant data storage is easily transferred from the backplane to the new signal converter. This technology equates to reduced process downtime to the user.

The OPTIBATCH 4011 is a specialized flowmeter engineered for use with linear and rotating filling machines. The 4011 maintains high accuracy in measuring with both conductive and non-conductive liquids and can be equipped with a selection of hygienic fittings. Accommodated line sizes range from ¼" to ½" in diameter.

The OPTIGAS 4010 series is another specialized flowmeter line engineered for use in compressed natural gas (CNG) dispenser applications. It is engineered as a twin V-tube design with industry-standard ¾" connections with adaptor options. The OPTIGAS is only available in a compact design.

DUAL TUBE METERS

Dual bent tube types are the workhorse of the Coriolis design configurations, primarily because this design allows the highest accuracy and greatest repeatability of measurement. These advantages are prized in many applications, of course, but especially in difficult ones such as material balancing, where different streams are being combined. Users strive to generate a process output that is very consistent but know that the incidence of measurement error can be multiplied throughout a system of measurement points. The high accuracy and repeatability of Coriolis flowmeters improve the likelihood that these errors will be kept to a minimum, increasing desired final product throughput.

BIBLIOGRAPHY

Flow Research, Inc., *The World Market for Coriolis Flowmeters*, 8th Edition (Wakefield, Massachusetts: Flow Research, Inc., 2025)

8 Thermal Flowmeters

OVERVIEW

This chapter examines the history and development of thermal flowmeters and their principle of operation, including illustrations to bring the operating principle to life. Thermal flowmeters evolved out of the technology of hot-wire anemometers, so we begin with a discussion of the development of anemometers. We also share some of our interviews with the founders of the main companies involved in inventing and producing thermal flowmeters in the early days (and today): Fluid Components Int'l, Kurz Instruments, and Sierra Instruments.

ANEMOMETERS

Anemometers are used to measure wind speed and the motion of air and some gases. The word "anemometer" came into use in the 18th century. It is a combination of the Greek word *anemos* for "wind" + meter, forming the duonym "windmeter." It means "an instrument for measuring the force, speed, and sometimes direction of the wind."

There are three main types of anemometers:

Hemispherical cup anemometers usually have three or four hemispherical cups on horizontal arms mounted on a vertical shaft. A simple instrument of this kind was developed in 1845 by Rev. Dr. John Thomas Romney Robinson. The flow past the cups is roughly proportional to wind speed at different velocities, so counting the number of turns by the shaft gives an approximation of the wind speed, taking into account turbulence produced by the windmeter, friction on the shaft, and other factors. The design was further modified by John Patterson in 1926 to produce a three-cup anemometer. Patterson modified the design of the cups to improve accuracy and to produce maximum torque. In 1991, Dr. Derek Weston further modified the design by attaching a tag to one of the cups. This makes it possible to measure both wind speed and wind direction (Figures 8.1 and 8.2).

In addition to the sensing element just described, a hemispherical anemometer also requires a transmitter or some device for counting the revolutions and converting this information into a reading of wind speed. Today's anemometers typically have a transmitter attached that counts the revolutions of the cups and also displays the windspeed, often in digital format. Software inside the transmitter can also make adjustments for friction and other variables that affect instrument accuracy and calculate wind direction when possible.

Vane anemometers work somewhat like hemispherical cup anemometers except that they have vanes instead of cups to capture the wind velocity. A vane is a flat blade. A vane anemometer works on the principle that a freely spinning turbine or rotor rotates at a speed that is directly proportional to wind speed. The turbine is also referred to as a propeller, impeller, or fan. Vane anemometers have variously shaped blades.

 DOI: 10.1201/9781003457855-8

FIGURE 8.1 An anemometer.

The origin of the vane anemometer can be traced back to the Italian artist and architect Leon Battista Alberti. Alberti described the theory of a mechanical device to measure wind velocity. A disc placed perpendicular to the wind direction caused it to spin. Wind velocity was indicated by the angle of the disc. In the 1640s, Evangelista Torricelli

FIGURE 8.2 A vane anemometer.

invented the barometer to measure air pressure. Various designs have been implemented in the centuries since then. In the 1930s, the first electronic anemometer was invented.

Like the hemispheric cup anemometer, a vane anemometer requires a transmitter or similar device to count the revolutions of the spinning blades and to compute the wind velocity. Today's instruments typically display the wind speed in digital format and are combined with other measurements such as wind direction, temperature, and barometric pressure.

Hot-wire anemometers are based on the principle that a heated wire cools as wind passes over it. A hot-wire anemometer consists of two probes with a thin wire stretched between them. The anemometer heats the wire to a known set temperature. As wind passes over the wire, it begins to cool. The anemometer calculates the wind speed based on the heat loss of the wire.

The first use of the term "hot-wire anemometer" is credited to Louis Vesso King in 1914. King formulated what has come to be known as King's Law, which mathematically describes heat transfer in air flows using a heated wire. As air moves over the wire, it causes a loss of temperature in the wire and removes some of the heat energy from the wire.

Hot-wire anemometers typically consist of two sensors. The first sensor is a temperature sensor that measures the temperature of the air passing over the sensor. The second sensor is a velocity sensor. The velocity sensor is heated to a temperature that is greater than the air temperature. As the air flows over the velocity sensor, it cools down the heated sensor. The cooling effect of the air on the heated sensor can be correlated to the velocity of the air.

Hot-wire anemometers were used in the 20th century for turbulence research, and to measure airflow at low velocities. An important step forward occurred when TSI undertook a project to design hot-wire anemometers for low-flow applications. TSI was founded in 1960 when it changed its name from "Pentalent Engineering Company" to "Thermo-Systems Engineering Company." In 1976, it changed its name to TSI, Incorporated.

In 1971, TSI took on a project to design a wind speed velocity sensor that is sensitive to low winds and winds in a low-density atmosphere. The company designed 60 anemometers for NASA, six of which went to Mars on the Viking spacecraft. TSI became one of the major suppliers of hot-wire anemometers, which were mainly used for turbulence research and for measuring air in low-flow applications.

FROM HOT-WIRE ANEMOMETERS TO THERMAL FLOWMETERS

The roots of thermal flowmeters go back to the hot-wire anemometers that were used for airflow measurement in the early 1900s. Hot-wire anemometers were used in velocity profile and turbulence research. They are very small and fragile, and consist of a heated, thin wire element. Hot wire anemometers have a quick response time because they are so small and thin. However, their fragility makes them unsuitable for industrial environments.

Thermal flowmeters were first introduced for industrial applications in the 1970s. The story of how they came on the market is a fascinating one that involves Sierra Instruments, Fluid Components Int'l (FCI), and Kurz Instruments. Sierra

Instruments and Kurz approached the subject through hot wire anemometers. FCI approached the subject through flow switches. All three companies were pioneers in the development of thermal flowmeters, and all three companies still offer thermal flowmeters today.

INTERVIEW WITH JOHN OLIN

The following is an excerpt from an interview done by Jesse Yoder with Dr. John Olin, founder of Sierra Instruments. The interview was done onsite at Sierra in November 2003. To see the complete interview, go to:

www.worldflow.com

This site is the Flow Research Knowledge Base.

Here JO stands for John Olin and JY stands for Jesse Yoder.

JO: As I said before, we started Sierra Instruments, Inc., in 1973. In 1975, we moved from Minnesota to the Monterey, California area. We put the business in two trucks, drove them over the continental divide, and ended up in this beautiful place where we wanted to spend our lives. In 1977, Jerry and I had an industrial divorce. Basically, we both wanted to go our separate ways and own our own companies. Each of us owned 50% of Sierra, so neither one of us had ownership control. We both had issues, and we settled out of court in 1977. I kept the name "Sierra Instruments." He needed a name: so, he selected "Kurz." He kept the flow instruments, and I kept the particulates instruments.

Now, I owned 100% of Sierra and was on my own. It was either sink or swim – failure wasn't an option. I gave it everything I had, and I loved it. I had no engineers working for me, so it all fell on my shoulders. I got very active in developing new products. I invented an instrument called the dichotomous sampler, which ultimately was specified by EPA and became one of their standards for air quality monitoring of airborne particulates. This instrument divided airborne particulates into two fractions – the respirable fraction and the non-respirable fraction – both of keen interest in the air pollution control field. In 1983, I sold my particulates instruments division to Anderson Samplers, a competitor, and got back into the flowmeter market.

JY: How did you get back in?

JO: For the first few years of our court settlement, Jerry Kurz and I could not compete. When the non-compete term was over, and I had sold my particulates business, I went back into the very thing I knew, which was flow.

JY: Did you invent a different product?

JO: When we split up, I kept my patents on particulates samplers, and Jerry kept one for flow. The only exception was a patented flow controller that used a thermal mass flow sensor but was used to maintain a constant sampling flowrate in a high volume air sampler. The technology fell into Jerry's corner, but the application fell into mine; so, our split-up agreement had Kurz manufacturing the product and Sierra, that is, me, selling it.

KURZ INSTRUMENTS THE FIRST TO BRING THERMAL FLOWMETERS TO MARKET

John Olin's account of the "industrial divorce" between Kurz and Olin is significant here because it enables us to definitively say which company first introduced thermal flowmeters. While it seems clear that John Olin and Jerry Kurz share the credit for inventing industrial immersion thermal flowmeters, it seems equally clear that Kurz Instruments was the first company to bring them onto the market.

Kurz and Olin moved from Minnesota to California in 1975 for the purpose of establishing Sierra Instruments there. Bringing a new product to market would require office and manufacturing space, potentially some approvals, and a path to market that includes sales channels. There is no indication that events were moving that quickly prior to the division of the company into two in 1977. While John Olin kept the name "Sierra," he did not keep the flowmeters, which went to Jerry Kurz, who established Kurz Instruments in April 1977. What John Olin kept was the particulates business. The key point is that John Olin had a non-compete clause that kept him out of the flowmeter business for a number of years, beginning in 1977. He returned to the flowmeter business in 1983. In the meantime, Kurz Instruments was free to sell the thermal flowmeters that Kurz and Olin had created together.

FLUID COMPONENTS INT'L (FCI) TOOK A DIFFERENT APPROACH

While the above discussion appears to show that Kurz and Olin together invented thermal flowmeters by taking the measurement principles used by TSI for hot-wire anemometers and developed an industrially hardened version of them that qualified as thermal flowmeters. As John Olin says:

> I mainly worked on thermal mass flowmeters. They made a thermal-anemometer probe – a probe with a hot tungsten wire on the end of it. You can measure free air flows in a room using a probe. You take the same probe and insert it in a pipe, and now it becomes a mass thermal flowmeter.

FCI took a different approach to developing thermal flowmeters. FCI was founded in 1964 by Mac McQueen and Bob Deane (Figure 8.3). The company developed flow switches to detect the flow of oil through pipes in the oil patch. Although these thermal switches were not flowmeters, they formed the basis of what later became flowmeters. In 1981, FCI put more sophisticated electronics on its switches, creating thermal flowmeters for gas flow measurement.

To read two interviews conducted by Jesse Yoder with Bob Deane and MacMcQueen conducted onsite at FCI in February 2003, go to the Flow Research Knowledge Base:

www.worldflow.com

In these interviews, Deane and McQueen recount the development of FCI from its founding in 1964 through to 2003.

FIGURE 8.3 Bob Deane and Mac McQueen.

THE SAGA CONTINUES

After FCI, Sierra, and Kurz got thermal flowmeter technology going in the 1970s, a second generation of suppliers appeared in the 1980s and 1990s. Eldridge Products, founded by Mark Eldridge, introduced its first thermal mass flowmeter in 1988. In 1993, Magnetrol brought out a thermal mass flowmeter. Fox Thermal Instrument, founded by Brad Lesko, was formed in 1994. Then, in 2002, Bob Steinberg formed Sage Metering. All these companies were in the Monterey, California area, with the exception of Magnetrol.

HOW THEY WORK

There are two different thermal flowmeter technologies. Some measure the speed with which heat added to the flowstream disperses. Others measure the temperature difference between a heated sensor and the ambient flowstream. Thermal flowmeters typically require one or more temperature sensors to measure the fluid temperature at specific points.

Thermal flowmeters work by introducing heat into the flowstream and measuring how much heat dissipates, using one or more temperature sensors. There are two different methods for doing this:

> One method is called the constant temperature differential. Thermal flowmeters that use this method have two temperature sensors. One is a heated sensor, and the other sensor measures the temperature of the gas. Mass flowrate is calculated based on

how much electrical power is required to maintain a constant temperature difference between the two temperature sensors.

The second method is called a constant current method. Under this method, thermal flowmeters also have two sensors; one heated sensor and one that measures the temperature of the flowstream. Power to the heated sensor is kept constant. Mass flow is measured based on the difference between the temperature of the heated sensor and the temperature of the flowstream.

Both methods make use of the principle that higher velocity flows result in greater cooling. Both compute mass flow by measuring the effects of cooling on the flowstream.

LATER DEVELOPMENTS

Fluid Components, Kurz, and Sierra were the main companies manufacturing thermal flowmeters in the early 1980s. However, soon other companies came along. In 1988, Mark Eldridge founded Eldridge Products in Monterey, California, a company that focused on developing a variety of types of thermal sensors to measure the flow of various types of gases. Eldridge developed a multipoint thermal flowmeter that is designed to be used in large cross-sectional areas where two or more measuring points are required. Examples are large intake ducts, air exhaust, and flue stacks. In the early 1990s, Eldridge's multipoint flowmeters were used for continuous emissions monitoring.

In 1994, Brad Lesko founded Fox Thermal Instruments in Gilroy, California, which is also in Silicon Valley. Fox Thermal's flowmeters are used for a wide variety of applications, including boiler inlet, compressed air, submetering and custody transfer, digester gas, wastewater aeration, and flare headers. Fox distinguishes itself with research into thermal sensors and manufactures its own thermal sensors (Figure 8.4).

In 2002, Bob Steinberg founded Sage Metering, after having worked for several other thermal flowmeter companies. Sage was founded in Monterey, California, where it is still located. Like the other thermal flowmeter companies described here, Sage focuses on gas flow measurement. However, Sage has developed a special focus on environmental applications, including measuring greenhouse gas emissions, flare gas measurement, carbon credits, and atmospheric storage tank vents.

OTHER THERMAL FLOWMETER COMPANIES

Thermal flowmeters are often associated with Monterey, California, and the surrounding area because of the concentration of thermal flowmeter companies there. However, there are a number of other thermal flowmeter companies in the United States and around the world. Thermal Instrument Company, located in Trevose, Pennsylvania, introduced thermal flowmeter technology to the commercial marketplace after 2000. Thermal Instrument produces both inline and insertion meters, and it produces them to measure flow in both liquids and gases. Another US company manufacturing thermal flowmeters is Magnetrol.

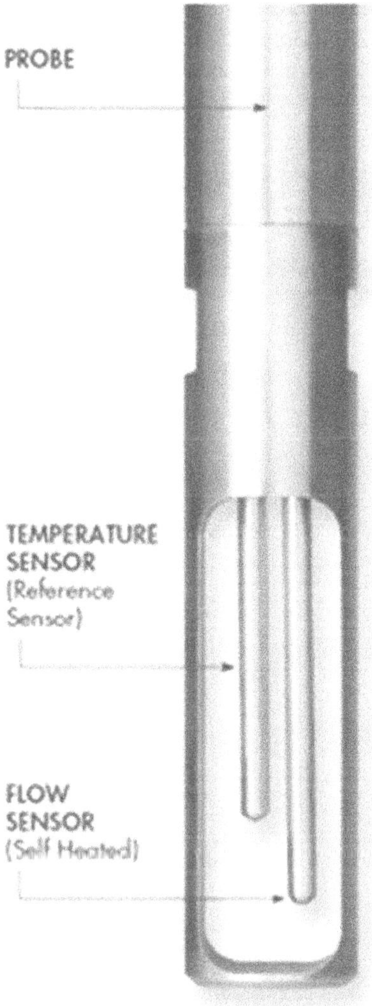

FIGURE 8.4 Example of dual temperature sensors to measure flowrate.

Source: Graphic courtesy of Sage Metering.

In Europe, both Endress+Hauser and ABB are major suppliers of thermal flowmeters. Unlike most of the US companies described above, Endress+Hauser and ABB offer a wide range of instrumentation products. These include multiple types of flowmeters, pressure transmitters, temperature transmitters and sensors, and analytical instruments. Both these companies are much larger in terms of total company revenues and number of employees than the more thermal-centric US companies.

In Japan, Tokyo Keiso is a major supplier of thermal flowmeters. Tokyo Keiso introduced its first thermal flowmeter in 1977, its first thermal mass flowmeter in 1983, and its mini-thermal flowmeter in 1991. The company's thermal flowmeter line is focused on the measurement of gases. These flowmeters are used in waste incineration plants, in air conditioning, and in many other gas supply applications.

Note: The "How They Work" section is excerpted from Chapter 8 of *New-Technology Flowmeters, CRC Press, by Jesse Yoder, 2023.*

A SAMPLING OF PRODUCT IMAGES FROM MARKET-LEADING SUPPLIERS

The following pages show pictures of a few products from market-leading thermal flowmeter suppliers.

Product analysis of all profiled suppliers follows after (Figures 8.5–8.12).

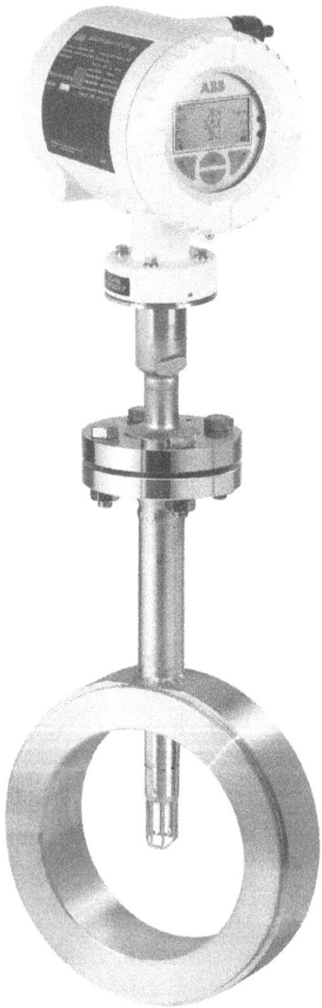

FIGURE 8.5 ABB's SensyMaster FMT400.

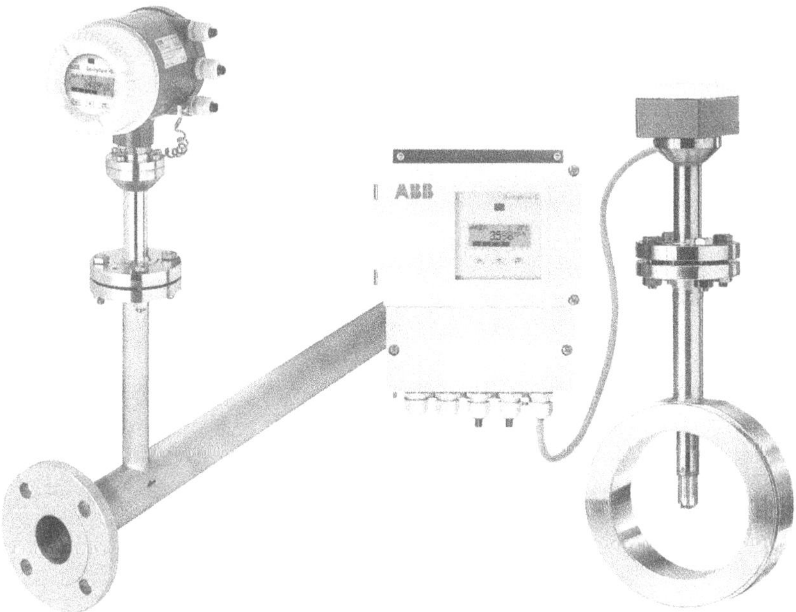

FIGURE 8.6 ABB's Sensyflow FMT500-IG digital thermal mass flowmeters.

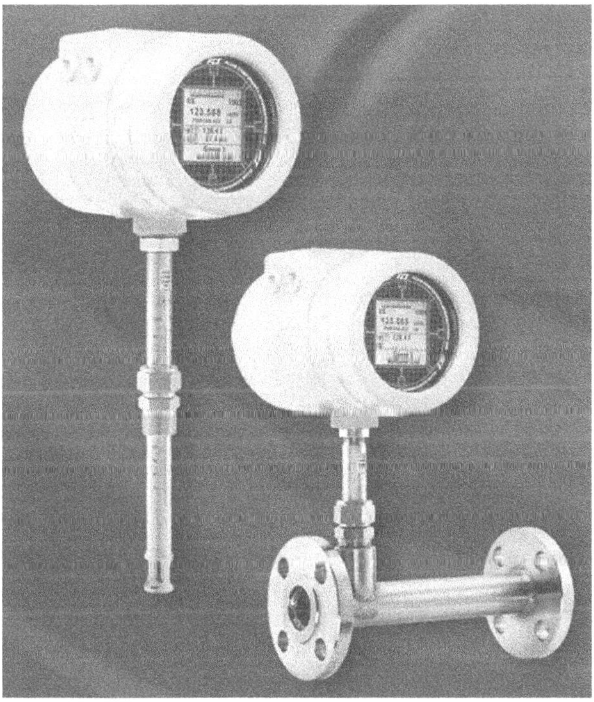

FIGURE 8.7 Fluid Components Int'l (FCI)'s ST100 Series air-gas thermal flowmeters.

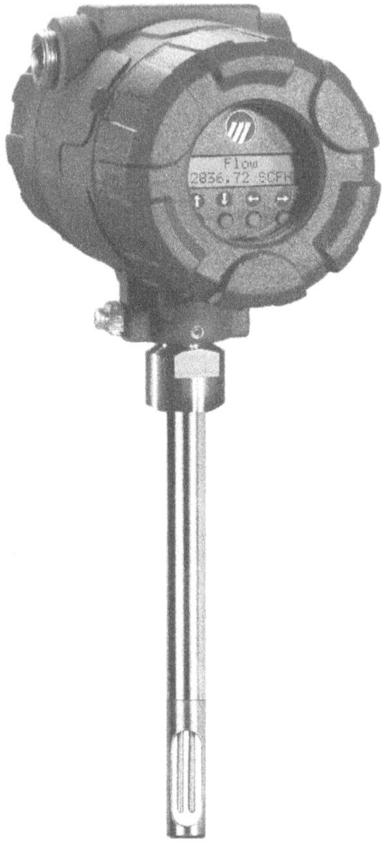

FIGURE 8.8 Thermatel® Model TA2 Thermal Mass Flow Meter AMETEK® Magnetrol USA, LLC.

Note: Magnetrol says its Thermatel® Model TA2 gas flowmeter is often utilized to manage energy costs or to meet environmental regulations.

INDUSTRIES AND APPLICATIONS

Thermal flowmeters, a cost-effective and versatile meter, are a mainstay for a wide range of applications and industries. Thermal flowmeters have fast response time, and they excel at measuring flow at low flowrates. They also provide a direct means of measuring mass flow and can handle some difficult-to-measure flows. The combination of value for money, high turndown ratio, relatively high accuracy in gases, cost-effectiveness in large line sizes, flexibility for a wide range of gases, and high repeatability accounts for a steady growth rate.

Note: In this chapter, we do not include mass flow controllers, which are in many cases a thermal flowmeter with an integrated valve that controls and measures flow. Instead, mass flow controllers are covered in Chapter 9 of this book.

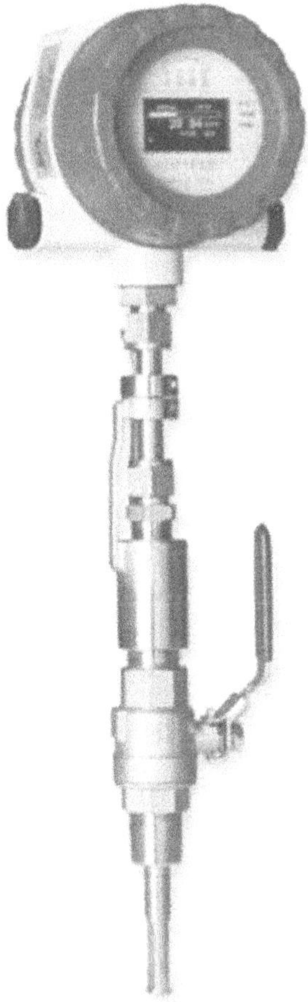

FIGURE 8.9 Sage's Prime is its flagship brand of gas thermal mass flowmeters.

Note: The integral version is shown here with its optional isolation valve assembly.

Insertion and inline thermal flowmeters both measure mass flow using heat transfer principles, but their design and installation differ significantly. Inline meters are installed directly into the pipe, using flanged or wafer connections. Insertion meters are inserted into the pipe and measure the flow at a single point, except for multipoint insertion meters.

The following is a comparison of their advantages and disadvantages.

Insertion Thermal Flowmeters

Advantages

Lower Installation Cost

- Easy to install on existing pipelines (hot tap option available in some cases).
- No need to cut the pipe or shut down the process.

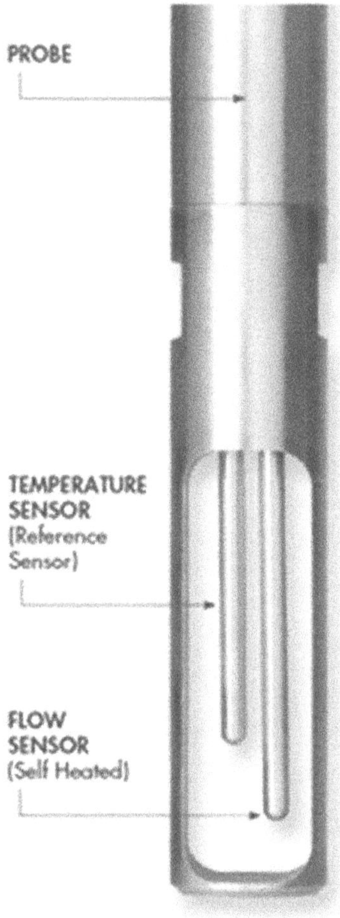

FIGURE 8.10 Sage dual sensors.

Minimal Pressure Drop

- Only a small probe intrudes into the flow, minimizing disturbance and energy loss.

Flexibility

- It can be moved between lines or relocated for temporary measurements.

Cost-Effective for Large Pipes

- Much less expensive than inline meters for large-diameter pipes.

Disadvantages

Lower Accuracy

- Typically, ±1–3% of reading, due to non-uniform velocity profile across the pipe.
- Insertion depth and location can affect readings.

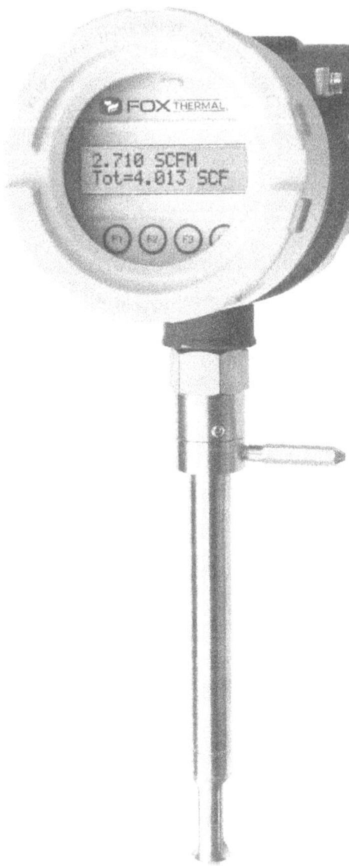

FIGURE 8.11 (TASI Group's) Fox Thermal's FT3 series also includes inline and remote models.

Vulnerability to Flow Profile Distortion
- Sensitive to swirl or poor straight-run conditions.

Mechanical Wear and Vibration
- Probe can be damaged in high-velocity or particulate-laden flows.

Limited Use in Small Pipes
- Not practical or accurate for small-diameter lines (under ~1.5 inches).

Inline Thermal Flowmeters

Advantages

Higher Accuracy
- Typically, ±1.0–3.0% of reading, due to controlled flow path and better sensor placement.

Stable Calibration
- Can be factory-calibrated as a system with pipe section included; more consistent performance.

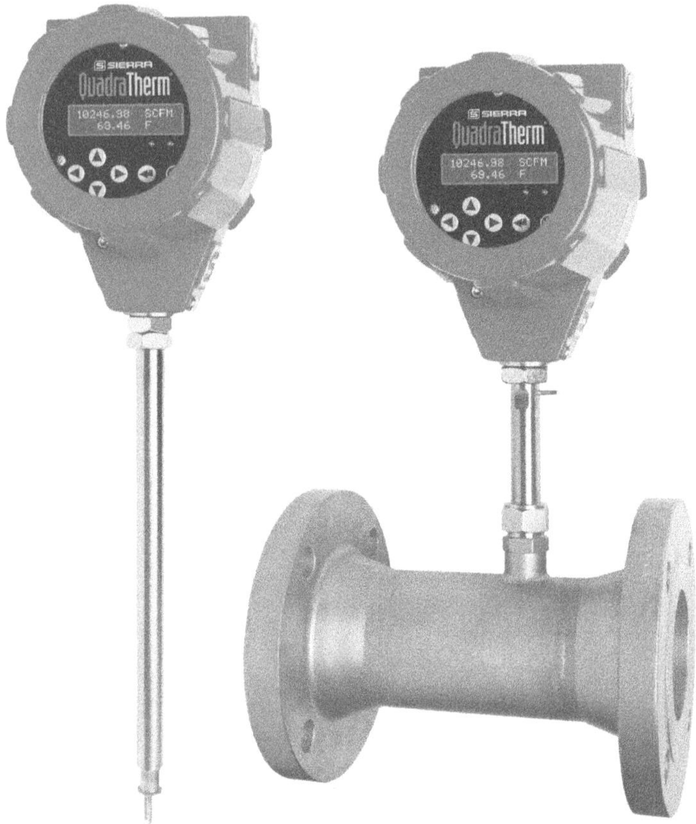

FIGURE 8.12 (TASI Group's) Sierra's QuadraTherm® 640i insertion mass flowmeter can be validated in the field for easy in-situ calibration. Sierra's QuadraTherm® 780i inline mass flowmeter reduces straight-run requirements with built-in flow conditioning.

Reliable for Low Flows

- Better sensitivity to very low velocities and wide turndown ratios (up to 1000:1).

Compact and Self-Contained

- Often includes built-in temperature and pressure compensation.

Disadvantages

Higher Installation Cost

- Requires cutting the pipe and interrupting flow to install.

Greater Pressure Drop

- Full-bore sensors can create minor restrictions in the line.

More Expensive for Large Pipes

- Cost increases significantly with pipe diameter.

Fixed Installation

- Not easily removable or transferable between lines.

MULTIPOINT INSERTION THERMAL FLOWMETERS

The difference between single-point and multipoint insertion thermal flowmeters lies in how many sensing elements are inserted into the pipe and how much of the flow profile they capture. The following is a comparison:

SINGLE-POINT INSERTION THERMAL FLOWMETERS

Description:

- Have one thermal sensing element inserted into the flowstream.
- Measure flow at a single point, typically in the center or at a fixed depth in the pipe.

Advantages:

- Lower cost and simpler installation.
- Suitable for small-diameter pipes or applications with uniform flow profiles.

Disadvantages:

- Limited accuracy in large pipes or ducts where velocity profiles vary.
- Prone to error in non-fully developed or skewed flow conditions.

MULTIPOINT INSERTION THERMAL FLOWMETERS

Description:

- Have multiple thermal sensors (e.g., 2, 4, 8, or more) along a rod or probe.
- Sample flow at different depths or radial positions, covering more of the flow profile.

Advantages:

- Better accuracy in large pipes, stacks, or ducts with non-uniform flow (e.g., elbows, bends, dampers upstream).
- Can provide average velocity or mass flowrate across the cross-section.

Disadvantages:

- Higher cost and more complex installation.
- May require longer straight pipe runs or custom mounting hardware.

THERMAL FLOWMETERS – YOU'LL FIND THEM EVERYWHERE

Thermal flowmeters are used in some of the world's most demanding processes and industries, including chemical, oil & gas, power & energy, water & wastewater, pharmaceutical, pulp & paper, nuclear power, and aerospace.

Although widely used, thermal flowmeters are still not readily accepted as a reliable measuring principle in all industries and markets. Due to the measuring principle itself, they are limited by flow profile dependency, point measurement, a strong dependency on pressure and temperature-related gas characteristics, the negative influence of condensed water, limited sensitivity, and strong temperature dependency in liquid applications.

One limitation of thermal flowmeters is that they are used almost entirely for gas flow measurement. Thermal flowmeters have difficulty in measuring liquid flows because of the slow response time involved in using the thermal principle on liquids. However, some companies now offer thermal flowmeters for liquid flow measurement. In this book, we identify the percent of thermal flowmeter revenues from liquid as opposed to air and gas applications as less than 2%. Clearly thermal flowmeters are used almost entirely for air and gas applications, and applications for liquids play only a minor role in the total market.

A second limitation is in their accuracy. Thermal flowmeters are not nearly as accurate as Coriolis meters, and typical accuracy levels are in the 1% to 3% range. However, thermal suppliers have been improving the accuracy of their flowmeters, with some even claiming as high as ±0.5% accuracy full scale under reference conditions for air measurement. As we discuss in the Frontiers of Research section later in this chapter, suppliers are also making product innovations that improve ease of use and accuracy.

The wide range of thermal flowmeter applications includes:

- Industrial and test gas rigs.
- Process industries with gas reactions and needs, including chemical, medical, pharmaceutical, oil & gas, food & beverage, automotive, general industrial markets.
- Air and gas compressors, compressed air systems, compressed air balancing, boiler air and gas fuel feed lines, natural gas submetering.
- Nuclear power plants.
- Environmental: flare gas, flare stack monitoring, vent gases, sewage gas, landfill gas, wastewater, aeration flow, digester gas, sewage treatment plants, measuring flue gas from industrial processes, air emission monitoring.
- Energy creation: synthesis gas, gasification plants, biogas, biogas in agricultural anaerobic digestion plants.
- Energy management, including natural gas and compressed air flow; measuring methane for commercial and industrial applications ranging from chemical production to advanced rocket fuels.
- Facilities management, including HVAC and building automation: submetering of gas usage, tenant billing, boiler efficiency, boiler/furnace gas flow, burners and boiler feed control, boiler inlet fuel flow, combustion air, drying air flow, vent air.

- Consumption and distribution of compressed air.
- Uncompressed air, nitrogen, and oxygen gas for a wide variety of commercial and industrial applications, from steel production (also argon) to dehydrating fruit to renewing municipal water supplies.
- Nitrogen blanketing in storage containers.
- Semiconductor gas supply lines, medical oxygen supply facilities, utility gas supply lines in industries, high temperature dry air lines, combustion gas lines, low pressure and low velocity gas lines, incineration facilities lines, electric power plant lines, flowrate measurement systems for automobile testing lines, and other gas supply applications.
- Pure gas applications.

FRONTIERS OF RESEARCH

Many thermal meter suppliers are introducing innovations and special-purpose products to meet specific industry needs and enhance functionality, versatility, accuracy, and reliability.

- Programmable for Changing Gas Needs.
- Versatility and Ease of Use.
- More Smart Meters.
- Greater Accuracy and Reliability.

PROGRAMMABLE FOR CHANGING GAS NEEDS

Suppliers are moving from meters that need to be factory calibrated for specific gases to thermal meters that customers can program to measure various gases and gas compositions.

FCI's Gas-SelectX® menu, for example, allows FT4A thermal meter users to choose from a preselected list of gases or create a custom gas mixture to measure with NIST-traceable precision. The software can validate calibration and print a certificate.

E+H offers meters with up to 22 selectable gases; two gas mixtures with up to 8 user definable components; toggling between two gas mixtures; and programmed gases that can be changed at any time without recalibration. The Proline t-mass transmitter's integrated "Gas Engine" function and automatic temperature and pressure compensation capabilities allow reliable measurement as gases change volume and properties due to pressure and temperature changes – without manually needing to convert operating volume to standard volume.

ABB's SensyMaster FMT250 features an Application Selector with up to eight applications and includes up to four customer-set applications, selectable gas type, and a selectable pipe inner diameter.

Sierra's Dial-A-Pipe and Dial-A-Fluid features allow users to change the pipe size and fluid type in the field without loss of accuracy. The company's qMix gas mixing software allows users to create custom gas mixtures in the field to compensate for known gas compositional changes, without recalibrating the meter.

VERSATILITY AND EASE OF USE

Suppliers are also including features that make life a little easier for thermal flowmeter users.

E+H's multivariable thermal meters provide direct measurement and display of both mass flow and fluid temperature, with no need for pressure or temperature compensation.

FCI's rugged, high-performance ST80 meter features an extensive selection of application-matched flow sensors, including FCI's patented "FPC" sensor for compressed air and a new "wet gas" flow element. The ST80 features Adaptive Sensing Technology (AST) and a patented hybrid sensor drive that combines both constant power (CP) and constant temperature (CT) thermal dispersion sensing technologies in the same instrument.

As Binder's COMBIMASS measures the flowrate; the meter also records and balances compressed air, technical and combustion gases such as nitrogen, argon, oxygen, carbon dioxide, superheated steam, propane, butane, and other gases.

Kurz has engineered a thermal mass flowmeter specifically for wet gas, condensing gas, and biogas environments.

E+H's thermal meters feature an optional "hot tap" mounting tool for inserting or removing the sensor during operation. Binder's COMBIMASS® hot tapping unit allows the flowmeter to be removed gastight during operations for checks and maintenance.

Magnetrol eliminates the need for a power backup by storing totalized data electronically. The company also offers a portable display model that permits multiple units to display their data at a single location.

Sage's energy-efficient thermal mass flowmeters feature a high-contrast graphical display of flowrates, temperature, total flow, a bar graph, and diagnostics.

Thermal Instrument's multipoint models combine up to six flow measurements and send them as one average flow to the electronic transmitter.

GREATER ACCURACY AND RELIABILITY

Through self-diagnostics, calibration, reducing the effects of disturbances, and other improvements, suppliers are delivering meters users can count on for greater accuracy and reliability.

Binder's COMBIMASS meters use multipoint sensors to reduce errors caused by flow profile distortions while providing the correct values of the flowrate even if the thermal flow sensors are not optimally positioned. Sensors measure the flow velocity of gases at various points of the cross-section and then average the individual readings. The company's patented flow conditioner helps create a reproducible flow profile for short straight pipe runs and high-accuracy applications.

Thermal Instrument flowmeters experience negligible pressure drop through a design that protects flow and temperature sensors from passing fluids at all times. The company's meters also feature redundant flow and temperature RTD sensors. Built-in alarms alert operators to issues or flow disruptions.

The sensors in Eldridge Products Inc.'s Master-Touch meters constantly measure the gas temperature and maintain a reference resistance on one side of a bridge circuit. Self-heating forces the second sensor, which is controlled by the temperature sensor and bridge, to a constant temperature higher than the gas stream temperature. EPI's bridge uses proprietary temperature compensation techniques to maintain the higher temperature and counterbalance the temperature effects.

FCI's patented Vortab Insertion Panel (VIP) flow conditioner neutralizes irregularities caused by elbows, valves, blowers, compressors, and other flow disturbances in piping and duct runs.

The inline Kurz 534FTB series incorporates patented technology to achieve low end-to-end pressure drop and fast response to velocity and temperature changes. Built-in inlet and outlet piping reducers/expanders minimize effects from upstream and downstream flow disturbances caused by elbows, valves, and line size changes. The company's constant-control and power-limiting design prevents sensors from overheating at zero flow. Continuous self-monitoring electronics verify the integrity of sensor wiring and measurements.

ABB's VeriMass onboard verification and diagnostics include fingerprint-based SensorCheck to verify the integrity of the thermal sensor elements, continuous health checks, and parameter plausibility checks.

Sierra offers long-term stability with a patented no-drift DrySense mass flow velocity sensor that provides in-situ calibration validation.

Magnetrol meters offer calibration verification in the field and automatic compensation for the temperature of the measured gas in formulating the flow measurement.

Sage builds flow conditioning into its inline flowmeters. It also digitally drives its sensors (instead of using a traditional Wheatstone Bridge circuit) to eliminate long-term drift and provide a reproducible zero flow data point. This data point in turn enables single-point in-situ calibration to verify that the flowmeter retains the original NIST traceable calibration, as well as that it is clean and hasn't drifted. The company's newest industrial thermal mass flowmeter, Sage Paramount™ features continuous diagnostics and fast response to rapid temperature fluctuations. Software allows customers to easily reconfigure meter settings (including changing the gas mix) and to use Bluetooth to run performance validation checks from a PC.

To protect the meter components, EPI's remote-style meters have two enclosures: one mounted at the point of measurement and a second remote enclosure that includes the electrical connections, linearizing electronics, and display/keypad assembly. A shielded cable carries the flow signal between the two enclosures.

One issue with manufacturer accuracy reports is that they are often given under ideal laboratory conditions and may not easily translate to the field. This issue applies to all flowmeter types, not just to thermal flowmeters. For insertion thermal flowmeters, because they are measuring flow at a point in the flowstream, it is difficult to get a highly accurate reading of mass flow. This is one reason that multipoint meters were developed, although they tend to be used on very large ducts and stacks. Also, in order to enhance accuracy, it is important to have the meter calibrated with the gas whose flow is being measured.

Some people say that thermal flowmeters are "a one percent meter," meaning that under normal conditions they can at best provide accuracy levels of 1%. This is not to say that under carefully controlled conditions a thermal meter might provide an accuracy level greater than 1%. Claims of higher than 1% accuracy should be examined carefully to make sure that all potential inaccuracies such as that of temperature and pressure sensors have been accounted for. It is also important to understand to what extent the quoted accuracy can be extended to actual operating conditions.

MORE SMART METERS

In 2008, when Flow Research collected data for its first thermal flowmeter study, smart thermal flowmeters accounted for about 20% of worldwide revenues. By the second study in 2016, this percentage had increased to 48%. In 2022, the percent of smart thermal meters was 68%. Clearly networking via communication protocols – and now, wireless – has become much more important over the past 15 years. Fox Thermal, for example, has introduced software that allows users to connect to and configure its FT4A thermal meter from a PC or laptop and, as mentioned earlier, Sage allows customers to use Bluetooth to run performance validation checks from a PC more than 50 feet away.

Some suppliers are also offering interoperability with automation systems so that they can be easily introduced within existing user networks. FCI, for instance, touts its verified interoperability with Emerson Process (Delta V), Yokogawa, and ABB (800xA) automation systems.

In terms of the communication protocols, HART has increased by about 10% in worldwide revenues for smart communication protocols since 2016. Another important change is that while Modbus remains popular, it has decreased significantly as a percentage of total smart communication protocol revenues in the same period. While the percentage of smart thermal flowmeters is somewhat less than that of other new-technology flowmeters such as Coriolis and ultrasonic, the trend is clearly toward producing more smart thermal flowmeters.

BIBLIOGRAPHY

Flow Research, Inc., *The World Market for Thermal Flowmeters*, 3rd Edition (Wakefield, Massachusetts: Flow Research, Inc., 2025)

9 Mass Flow Controllers

OVERVIEW

Anyone who follows the flowmeter business is likely to know that many mass flow controllers (MFCs) also employ thermal technology. Should these devices also be included in any discussion of the history of thermal flowmeters, or should they be treated separately? As it turns out, the history of MFCs is also very fascinating, and MFCs developed independently and somewhat in parallel to thermal dispersion flowmeters. MFCs are also a particularly attractive solution for many OEM and System Integrator applications where air, oxygen, and/or hydrogen flow is required.

HISTORY

In the fall of 1954, someone from Exxon approached Charlie Hawk, an engineer from Hastings Instrument Company (now Teledyne Hastings), at a trade show. He wanted an instrument that would measure the amount of waste gas flowing in a flare stack so he could better monitor how efficient his refining process was.

HASTINGS INTRODUCES AIR FLOW PROBE IN 1955

In response to this inquiry, Hastings introduced what it called an "air flow probe" in 1955. This device contained small thermocouples in the flowstream, which measured the rate of cooling. Unfortunately, the Hastings air flow probe became covered with grime and could not survive the heat in this application. As a result, the instrument was withdrawn after several years.

Capillary Tubes Introduced by Hastings in the 1970s

In the early 1970s, Charlie Hawk and his colleagues decided to try again. Instead of putting the air flow probe directly into the flare stack, they put the probe into a capillary tube that had the sensors on the outside. This is essentially the principle that a mass flow controller uses to measure flow. This new device was more successful. While the tubes still became plugged, it was possible to clean them by purging them with a source of air.

This brief history shows that important developments in thermal metering were in fact occurring at about the same time that thermal flowmeters were being developed on the West Coast of the United States. In many ways the experiences were similar. The Hastings air flow probe, introduced in 1955, had to be withdrawn because it was not rugged enough for industrial flare stacks. This sounds very much like the anemometers from TSI in Minnesota that were used for research purposes. They were not industrial thermal flowmeters because they were not rugged enough, although

DOI: 10.1201/9781003457855-9

industrial thermal flowmeters were developed from them. So, if we consider the Hastings product a thermal flowmeter, then TSI's probes would have to be similarly classified.

THERMAL FLOWMETERS AND MFCs TAKE INDEPENDENT TRACKS

The Hastings instruments developed in the 1970s may have used a thermal principle, but the design is that of an MFC. Because industrial thermal dispersion flowmeters and MFCs are so different in type and application, it is probably best to consider them as independent instrument types and to describe their histories separately.

THE ROOTS OF MASS FLOW CONTROLLERS

Some of the earliest ideas for a thermal MFC go back as far as 1911, when C.C. Thomas described a flowmeter containing a heated element with two thermometers upstream and downstream of the heated element. However, this instrument was never put into production. In 1947, J.H. Laub proposed the idea of attaching the heater and the thermometers to the outside of the pipe. While this instrument seemed to give good results under ideal conditions, its readings became unstable when faced with the realities of gas viscosity, Reynolds number, and velocity profile.

Like industrial thermal dispersion flowmeters, MFCs for industrial use appear to have developed mainly in the 1970s, although some sources credit Tylan General (now part of Millipore) with introducing capillary tubes in the 1960s. Besides the advances made by Charlie Hawk and others at Hastings, others developed MFC technology independently during this time. One development involved a micromachined silicon sensor with a very fast response time.

THERMAL MFC PRINCIPLE OF OPERATION

MFCs both measure the mass flow and control it. MFCs have a proportional control valve that opens or closes to allow or stop the flow based on a setpoint or desired value.

Most MFCs utilize a bypass that diverts some of the flow into a smaller capillary tube. This is the method used by thermal MFCs, which are the dominant MFC technology. This method is described as follows.

Main Components

1. **Flow body** – where the gas flows through.
2. **Bypass and sensor tube** – splits off a portion of the gas for measurement.
3. **Thermal sensors** – placed upstream and downstream of a heating element on the sensor tube.
4. **Heater** – heats the gas in the sensor tube.
5. **Control valve** – adjusts the flow rate based on feedback.
6. **Electronics/controller** – compares actual flow to setpoint and controls the valve.

Working Principle: Thermal Mass Flow Measurement

1. **A small portion of the gas** is diverted through a **sensor tube** parallel with the main flow (a laminar bypass).
2. A **heater** warms the gas in this tube.
3. **Temperature sensors** on either side of the heater detect the heat carried by the gas.
4. The **difference in temperature** between upstream and downstream sensors is proportional to the **mass flowrate** (since more mass carries away more heat).
5. The electronics convert this temperature difference into a **mass flow signal**.

Control Function

1. The user sets a **desired mass flowrate** (the **setpoint**) electronically.
2. The controller compares the **measured flow** to the **setpoint**.
3. It sends a signal to a **proportional control valve** to **open or close** slightly to maintain the desired flow.

The large majority of MFCs use a thermal measurement principle to measure the mass flow in the bypass. The mass flow controller measures the heat dissipation in the bypass capillary tube, using resistance temperature sensors, and uses this value to compute mass flow. The paradigm case application for thermal MFCs is the same as for thermal flowmeters: **clean, pure gases whose heat capacity is known, or mixtures of gases, with known composition.**

While some of the newer technology MFCs eliminate the need for a bypass, many MFCs still typically have a bypass design, with some of the fluid being diverted to pass through a small capillary tube. The flow passing through the small capillary tube is measured and this value is used to compute the total mass flow passing through the flowmeter.

In some cases, the flowmeter has a bypass for measurement purposes but does not have a control valve. In this case, the flowmeter can measure the flow but not control it. These products are called mass flowmeters, rather than MFCs. MFCs are also differentiated according to whether they use thermal, differential pressure (DP), Coriolis, or ultrasonic technology to compute flow.

CORIOLIS MFC PRINCIPLE OF OPERATION

The Coriolis lines of flowmeters and controllers are of particular interest. Compared to traditional thermal MFCs, Coriolis-based flow controllers offer better accuracy and operate independent of fluid properties. Further, the Coriolis sensor is by nature faster than a sensor based on heat transfer. The ultrasonic technology MFCs also offer faster response time to changes in flowrate than their thermal versions.

Unlike thermal MFCs (which split off flow through a bypass sensor tube) or laminar DP meters (which have flow go straight through a restriction), Coriolis meters work differently. Coriolis meters do not need to go through a bypass capillary tube. Coriolis flowmeters, as MFCs, work much the same way that they work in other applications. As we learned in Chapter 6, Coriolis meters work in the following way.

How Coriolis Meters Work in a Mass Flow Controller

1. The entire fluid stream flows through one or more oscillating tubes.
2. An electromagnetic drive coil causes the tube(s) to oscillate at their natural frequency.
3. Pickoff sensors (mounted at the inlet and outlet of the vibrating section) detect the motion of the tubes.
4. When fluid is not flowing, the tubes oscillate symmetrically.
5. When fluid is flowing, its inertia causes the tubes to twist slightly – creating a phase shift between inlet and outlet sensor signals.
6. This phase difference is directly proportional to mass flowrate.

Why There's No Bypass

- The inertial response that creates the measurable twist requires all the mass to flow through the oscillating tubes.
- If any flow were bypassed, it wouldn't be subject to the oscillatory acceleration, so it wouldn't contribute to the measured signal.
- That's why Coriolis meters measure the entire flow directly.

DP LAMINAR FLOW MFCs

A DP laminar flow MFC works by measuring the pressure drop across a laminar flow element (LFE) and using that to calculate the mass flowrate of a gas. It is commonly used in gas delivery systems in industries like semiconductors and analytical instrumentation.

BASIC PRINCIPLE

The working principle is based on laminar flow and Poiseuille's Law, which states that for laminar (non-turbulent) flow, the pressure drop (ΔP) across a restriction is proportional to the volumetric flowrate (Q_v). Laminar flow MFCs do not use a flow bypass. Instead, they use a series of LFEs to create the laminar (non-turbulent) flow.

A LFE typically consists of:

- A bundle of small capillary tubes, or
- A porous sintered material, or
- Parallel plates with narrow gaps

Figure 9.1 depicts a mass flow controller from Alicat Scientific using an LFE to measure gas flow.

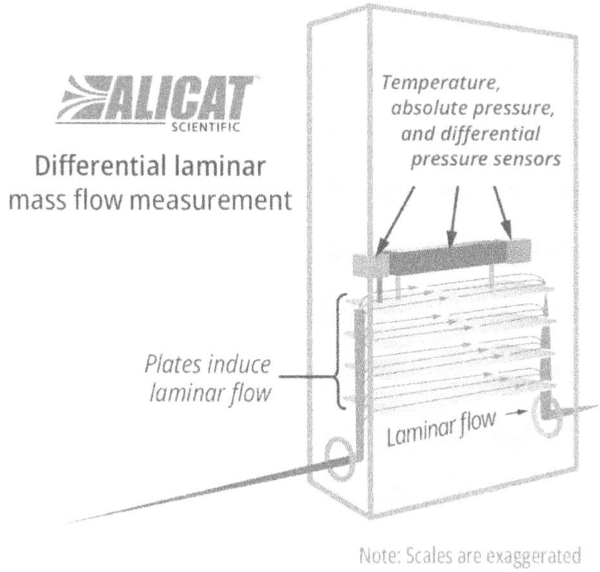

FIGURE 9.1 An illustration of an Alicat Scientific mass flow controller with DP technology. Used with permission of Alicat Scientific, Inc.

MOUNTING TYPES

There are a number of ways in which thermal flowmeters differ from MFCs. One way is in mounting types. MFCs typically are made as inline devices, fitting into a relatively small gas flow line. Thermal dispersion flowmeters, by contrast, are divided between inline and insertion meters. About 65% of thermal dispersion flowmeters are of the insertion type, while the remaining 35% are inline. The bulk of these are single-point insertions, while the remaining types are multipoint insertions. Insertion thermal flowmeters can be used on large line sizes, including those above 20 inches.

INDUSTRIES AND APPLICATIONS

Another important difference between MFCs and thermal dispersion flowmeters is that a majority of MFCs are used in the semiconductor industry. The remaining MFCs are divided between industrial and research/laboratory applications. The industrial applications are divided among a variety of industries and applications. Some of these include petrochemical, metals processing, food and beverage, solar/photovoltaic, fuel cells, and a number of others. MFCs are widely used to measure industrial gases, whether they are in semiconductor, industrial, or laboratory/research applications.

Thermal flowmeters, by contrast, are mainly used in the process industries. Leading industries for thermal flowmeters include power, water and wastewater, chemical, and oil and gas. Thermal flowmeters are used almost entirely for gas flow measurement in those industries, and they are used both to measure natural gas and industrial gases. Many thermal flowmeters are used for environmental applications. These include continuous emissions monitoring (CEM), flare and stack gas measurement, biogas, biomass, compressed air, boiler inlet, and many others. Multipoint thermal flowmeters were developed to provide higher accuracy in larger line sizes such as those above 30 inches.

AN ATTRACTIVE SOLUTION FOR OEM'S AND SYSTEMS INTEGRATORS

The majority of thermal flowmeters are sold to end-users, with the balance being sold to OEMs, systems integrators, and engineering and consulting firms. However, a larger percentage of MFCs are sold to OEMs and systems integrators. MFC OEM applications include solar energy and semiconductor equipment manufacturing. Other examples include biotechnology and pharmaceutical applications. These four industries have a high requirement for measuring air, oxygen, and hydrogen flow. MFCs fit well for these applications because they are small and economical, yet sufficiently accurate for the jobs they are required to perform in these environments.

A particularly common application of MFCs for OEM applications is in the pharmaceutical industry for dosing applications. Banks of MFCs are used to fill 15 or more ampoules at a time. In this case, the OEM sets up the groups of MFCs specifically for the purpose of filling the ampoules. Both repeatability and accuracy are important. The OEM sets up the equipment, which includes multiple MFCs, so that it is ready to be put into the dosing line.

OEMs play a vital role in the flow measurement market. A flowmeter, for example, may be ready to go out of the box, but it may also require other instruments, such as temperature sensors, pressure transmitters, or analytical equipment to serve a particular application. An OEM purchases the flowmeter and incorporates it into an instrument assembly that is designed for a particular purpose. Another example is skids used in downstream custody-transfer applications for refined fuels. These typically incorporate positive displacement or Coriolis flowmeters, along with many other devices. A different example is that of orifice meter runs, which incorporate an orifice fitting along with upstream and downstream meter runs (Figure 9.2).

RESEARCH CONTINUES

It is fascinating to look back at the origins of different flow technologies and to come to understand who made the important contributions that have gotten us to where we are today. It is equally interesting to try to understand where these technologies are headed, and what the next breakthroughs will be.

In the case of MFCs, a lot of attention is being paid to the industrial segment – in part to compensate for the cyclical nature of the semiconductor industry. In the case of thermal flowmeters, suppliers are working to increase their accuracy and to target specific new environmental applications. Research continues in both technologies.

Research is also continuing in other existing technologies. In addition, new flowmeter technologies are currently being developed that could change the flowmeter playing field. Of course, any new technologies will have to go through the same development and adoption process of other flowmeter types. The important thing to

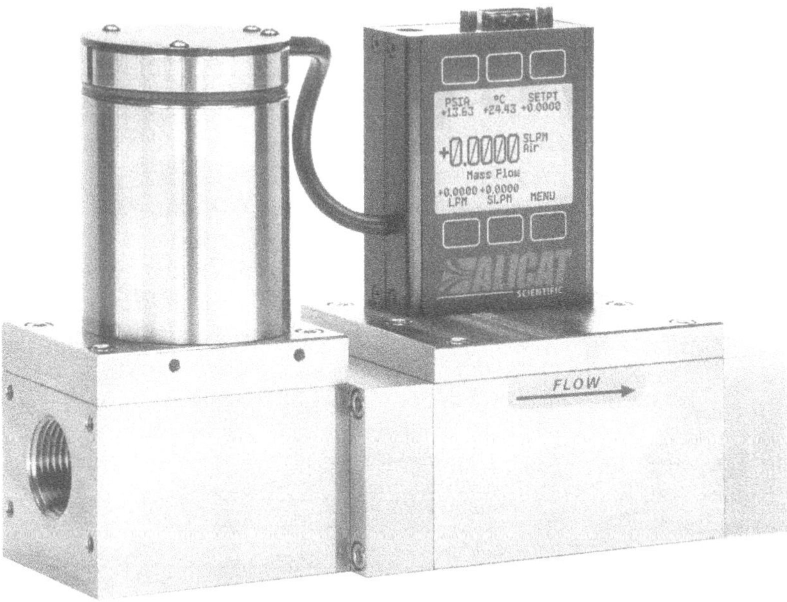

FIGURE 9.2 An Alicat Scientific mass flow controller. Used with permission of Alicat Scientific, Inc.

remember is that the flowmeter market is dynamic, not static, and that just as mergers and acquisitions can suddenly alter the landscape, so can the introduction of new flowmeter technologies.

THE SEMICONDUCTOR INDUSTRY WILL CONTINUE AS A MAJOR DRIVER OF MFC DEMAND

The semiconductor market accounted for almost 60% of the revenue for MFCs shipped in 2024. The semiconductor market is a cyclical market, and market cycles often last between 12 and 18 months. While the semiconductor market has experienced downturns in 2000–2001, 2009–2010, and again for the greater parts of 2011–2013 (along with the global economy), it has experienced recoveries just as robust recoveries in each instance. During the early part of the COVID-19 pandemic, the semiconductor market experienced a decline due to uncertainty and supply chain issues. By late 2020 and into 2021, the semiconductor industry experienced robust revenue growth, driven by surging demand across various sectors. According to the Semiconductor Industry Association (SIA), the global semiconductor industry sales increased by 3.3% in 2022, reaching a record high of $574 billion. However, the SIA also says that in 2023 the worldwide semiconductor industry declined by 8.2%, with total sales reaching $526.8 billion compared to the previous year's record high of $574.1 billion. In 2024, total semiconductor sales climbed to $627.6 billion, an increase of 19.1% over the figure in 2023. This demand is driven by strong demand for data center memory chips and memory products.

DEMAND FOR MASS FLOW CONTROLLERS WILL CONTINUE TO COME FROM THE EMERGING MARKETS, INCLUDING CHINA, INDIA, AND SOUTHEAST ASIA

The economies of China, India, and the Southeast Asian countries are continuing to grow at a rapid pace. Even if these economies were affected by a global slowdown, their continued population growth will result in demand growth in a variety of industrial segments including energy consumption, semiconductors, pharmaceuticals, food & beverage, power, and other segments. China and India will also be a focused source of continued demand for computers and semiconductor products, and the upgrades that consumers in other regions so often seek for these types of devices will develop into a robust secondary and replacement market here as well. All these forces will work towards increased use of MFCs in the future.

In the Middle East and Africa, expanding oil and gas infrastructure and water treatment plants are driving demands for MFCs. This is a global trend that will affect other regions as well.

MASS FLOW CONTROLLERS ARE REPLACING MANUAL AND NON-ELECTRONIC DEVICES

In some cases, gas flow is measured and controlled using a combination of a variable area flowmeter and needle valve. While some variable area flowmeters have been developed with an output signal, the majority of these flowmeters still need to be read manually. In many cases, the needle valve is adjusted manually. However, the trend is well along today toward the automation provided by MFCs.

This is a trend that is now well-recognized by end-users. End-users in general are looking for the higher accuracy of MFCs, and they are also looking to automate their gas measurement and control systems when this is possible. This is especially true in the most refined applications such as are associated with bioreactors and fermenters used in the life sciences industry. The careful measurement and control of CO_2 and O_2 gases in the manufacturing of cell cultures, for instance, are ideal applications for MFCs and their design promotes the trend toward process automation.

ADVANCES IN MFC TECHNOLOGY

The major manufacturers of MFCs, including Brooks Instrument, Bronkhorst, Horiba, and others, have made major advances in MFC technology. Developing MFCs with higher accuracy, real-time diagnostics, and multi-gas capability is driving market growth. Compact MFC designs enable use in smaller devices and tighter spaces in a drive toward miniaturization. Smart MFCs with connectivity features allow remote monitoring and predictive maintenance, enhancing their appeal. Other advances include introducing new communication protocols like EtherCAT and CANopen, and a number of other protocols.

Another important development is the further introduction of new flow technologies for use in MFCs. These include DP, Coriolis, and ultrasonic. While thermal is still the dominant technology, these alternative technologies provide a wider choice to users that they may choose, depending on their applications.

INDUSTRIAL APPLICATIONS FOR MASS FLOW CONTROLLERS

Table 9.1 gives an overview of the main industrial segments for mass flow controllers and lists examples of applications within them (Table 9.1).

TABLE 9.1
Industrial Applications for Mass Flow Controllers

Industrial Segment	Applications
Aerospace	Hydraulic systems test and fabrication; Ventilation R&D; Hardening canopies for jet aircraft; Spacesuit testing
Alternative Fuels	Measuring and controlling the use of hydrogen, biogas, ethanol, biodiesel, natural gas (CNG/LNG), and synthetic fuels; fuel cells; solar power; ethanol and biodiesel production
Analytical / Gas Analyzers	Analytical sampling; Gas sample preparation and measurement; Verifying flow and pressure for multiple gases flowing to and from gas chromatographs
Automotive	Emissions testing; Emissions monitoring; Measuring compressed air; Verification of SHED (Sealed Housing for Evaporative Determination) operations; Exhaust gases
Biotech / Pharmaceutical	Process control of reactor gases to fermentation; Bioreactor gas management; Maintain quality in drug manufacturing, in production, in FDA testing, and in culture growing
Chemical / Petrochem.	Measurement of gases in chemical processes and manufacturing
Electronics Manufacturing	Manufacture of computers, monitors, and other electronic equipment; Laser welding and cutting
Fiber Optics / Glass	Glass manufacturing; Ultraviolet coating on glass; Fiber optics and glass coating; Bulletproof glass for cars; High purity optics for bathroom faucet coatings
Food & Beverage	Blending, Process control in bottling, drying, mixing, cooling; Protective gases for packaging; Wine and beer making
Furnaces	Flame control; Gas mixing and blending; Burner control
Gas Distribution	Gas consumption measurement for internal accounting purposes
Heat Treating	Burner control; Welding
Lab Grown Diamonds	Control of gas flows, especially methane and hydrogen; Used in chemical vapor deposition (CVD) process to deposit carbon atoms onto a diamond seed substrate, forming a diamond layer
LED Lighting	Particulate dispersal, gas used in deposition, OLED
Medical	Check performance of equipment; Anesthesia; Medical equipment manufacturing
Metals Processing	Improve the quality of manufactured metals
Packaging	Protective gases for packages
Solar / Photovoltaic	Application of thin film coatings (to panels)
Power	Measurement of gases used in power generation
Thin Film Coating	Used in many industries including electronics, optics, solar panels, and medical devices to ensure that the gases involved in the coating process are delivered at accurate flowrates and ratios.
University / Government / Research lab	Research & Development; Testing; Calibration
Other	Fuel testing; Geothermal, other alternative energy & fuels R&D, production; Testing, adjusting motorized equipment (e.g., 2-stroke engines); Aeration for fish farming; Equipment & appliance testing

LEADING SUPPLIERS OF MASS FLOW CONTROLLERS

Horiba

HORIBA manufactures thermal and pressure-based MFCs for gas and liquid. All of its MFCs offer high accuracy and fast response times. The company also offers Coriolis flowmeters. HORIBA primarily sells semiconductor production equipment and device manufacturers. Primary applications are controlling the flow of gases and liquids, monitoring cleaning fluid concentrations in semiconductor manufacturing processes, and semiconductor quality control inspections. HORIBA's MFCs offer high accuracy, digital linearity, and the ability to use multiple calibration curves for different gases.

HORIBA developed the first mass-produced mass flow controller in 1990 and has continued to advance its designs with higher-performance central processing units and improved software. In 2023, the company introduced low-flow rate, pressure-type MFCs that support new communications specifications for handling next-generation semiconductor production processes.

HORIBA's CRITERION D500, designed for the most advanced semiconductor manufacturing processes, is a leading-edge, high-performance, pressure-insensitive mass flow module installed with DP detection. The multi-range/multi-gas/multi-pressure device includes the company's G-LIFE (Gas Law check of Integrated Flow restrictor Equation) self-diagnosis function for advanced process health monitoring, typically in three seconds or less. The new D700T pressure-insensitive mass flow module adapted the HORIBA CRITERION pressure-based flow measure and control technology to provide industry-leading repeatable and accurate performance.

HORIBA's thermal gas MFCs incorporate a flowrate measurement section with a sensor, bypass, flowrate control valve, and special circuitry. The company offers programmable multi-gas MFCs and configurable multi-range MFCs.

The company's best-selling SEC-Z500X Series was the firm's first line to reduce the number of MFCs required to cover the range of flows and gases used in semiconductor processes. Users can change the type of gas or full-scale flowrate on site quickly and easily.

The SEC-Z700S Series controls gas flow without effects from upstream and downstream pressure fluctuations. A newly developed thermal sensor and a pressure-insensitive function provide stable flowrate control. The series features on-board pressure measurement and provides readings for flowrate output, temperature, and pressure.

BROOKS INSTRUMENT

Brooks Instrument was founded by Stephen A. Brooks in 1946 in Lansdale, Pennsylvania as the Brooks Rotameter Company. Brooks later became the first instrumentation company in the world to be certified to ISO 9001 quality standards. Today, the company operates facilities in the US, Japan, the Netherlands, Singapore, Korea, and China. The firm maintains a worldwide sales and support network of representatives.

In April 2001, Brooks became a part of Emerson's group of flowmeter companies that included Rosemount, Daniel Industries, and Micro Motion. On December 31, 2007, Emerson sold Brooks to American Industrial Partners Fund IV for approximately $100 million. In May 2009, Brooks acquired Key Instruments, Inc., a variable area flowmeter manufacturer based in Trevose, Pennsylvania, offering acrylic and molded plastic, glass tube meters, and electronic flowmeters as well as flow control valves for medical, industrial, water, chemical, and laboratory applications. The Key Instruments product line has been absorbed within the Brooks portfolio.

In January 2012, American Industrial Partners sold Brooks to ITW. Brooks is now positioned within ITW's Test & Measurement and Electronics operating segment, which is one of the largest among ITW's seven segments. Brooks maintains offices and manufacturing sites in the US, Brazil, Europe, UAE, and Asia/Pacific.

MASS FLOW CONTROLLERS

Brooks Instrument has one of the widest MFC portfolios in the industry. It divides its thermal-based MFCs into three basic wetted-part categories: elastomer sealed, metal sealed, and high temperature. The elastomer and metal sealed versions are designed for gas measurement. The high temperature versions, which also have metal seals, are designed for measuring both vapors and gases. Brooks also offers flowmeters and MFCs for gas and liquids based on Coriolis technology.

Key applications for Brooks' MFCs include heat treating, cutting, and other thermal processes for mass flow sensors; semiconductor and LED production tools, thin film deposition systems; biopharmaceutical bioreactors; petrochemical and chemical gas and liquid pilot plants; and purge gas, process analysis, and analytical flow measurement.

Brooks' elastomer-sealed MFCs and meters for gas feature the widest flow and pressure ranges, long-term stability, products approved for hazardous areas, and hose down/wash down device enclosures.

Metal-sealed MFCs and meters are designed for processes sensitive to moisture or oxygen. They are noted for ultra-fast response times and a high-purity flow path that minimizes process contamination and enhances yield. The 5850EM series, which offers ultra-pure performance for high-pressure applications, has a large installed base with proven long-term performance.

The company's high-temperature group of MFCs and meters offers the highest operating temperature (up to 150°C), high accuracy and repeatability, and ultra-high purity construction. They are designed for optical fiber manufacturing (MCVD) and silicon semiconductor device fabrication processes. Brooks' new GF120xHT MFCs and meters, for instance, are designed for the most advanced semiconductor etch, thin film, and other gas process control applications. They control and measure the mass flow rate of vapor from liquid or solid sources to support the growing use of these precursors in semiconductor manufacturing processes.

The company's Coriolis-based Quantim MFCs and flowmeters for liquids and gases measure flow, density, and temperature. They provide the highest accuracy for very low-flow processes. Key applications for ultra-precise direct mass flow measurement include catalyst research, liquid precursor delivery to vaporizers, precision thin film coating and surface treatments, hot melt extrusion for pharmaceuticals, high-accuracy filling stations, and metering pump feedback.

BRONKHORST

Established in 1981, Bronkhorst has always upheld a strong culture rooted in family, loyalty, and sustainability. The company's dedication to quality is evident through its numerous certifications, including ISO 9001 for Quality Management and ISO 14001 for Environmental Management. In 2010, the Bronkhorst Calibration Centre received the highest international level of calibration security by being accredited to the International Standard ISO/IEC 17025.

In 1997, the acquisition of Manger + Wittmann Elektronische Gasmesstechnik GmbH, a German manufacturer of mass flowmeters and controllers, further strengthened Bronkhorst's capabilities and product portfolio. The company, later renamed Bronkhorst Instruments GmbH, specializes in the production of thermal mass flowmeters and controllers for gases, focusing on direct through-flow measurement. At the end of 2021, Bronkhorst Instruments relocated to a state-of-the-art facility in Leonhardsbuch, Germany, which includes a Class VII certified cleanroom for microchip and pharmaceutical flow instruments.

MASS FLOW CONTROLLERS

Bronkhorst manufactures Coriolis and thermal mass flowmeters and controllers for gases, liquids, and vapors and complementary lines of pressure and volumetric liquid flowmeters and controllers. The company offers a wide variety of both standard and custom designs for use in laboratories, machinery, industry, and hazardous locations. Bronkhorst claims to offer the widest product range of low-flow mass flowmeters and controllers on the market today. The company's mass flowmeters and controllers for liquids and gases are used in laboratories, test benches, machinery, and a wide variety of industries.

The company's thermal mass controllers for gas feature excellent accuracy, high repeatability, small footprint, an on-board gas conversion model, large gas database with customer selectable gases, and low cost of ownership. They are used in a variety of markets: analytical, chemical, energy transition, semiconductor, food & beverage, and surface treatment.

In 2022, Bronkhorst introduced a line of ultra-compact, chip-sensor-based mass flow instruments to measure and control gas flow and pressure. The innovative new FLEXI-FLOW Compact flowmeter and controller combines fast and stable chip sensors with reliable and accurate bypass technology. It measures flow, temperature, and pressure in one instrument and adapts to many applications through a wide dynamic range of measurement and control. The series features an onboard gas database and real-time conversion for superior accuracy.

Bronkhorst also specializes in liquid flowmeters, controllers, and dosing solutions for micro to low flow rates. The liquid meters measure multiple parameters – flow, temperature, density, and speed of sound– and can handle unknown fluids because they operate independent of fluid properties. They offer high dosing accuracy, onboard intelligence for fast and compact control loops, low cost, and no maintenance (no moving parts). Thermal mass flow instruments for liquids range from 5 mg/h to 1 kg/h; Coriolis mass flow instruments range from 50 mg/h to 600 kg/h (Figure 9.3).

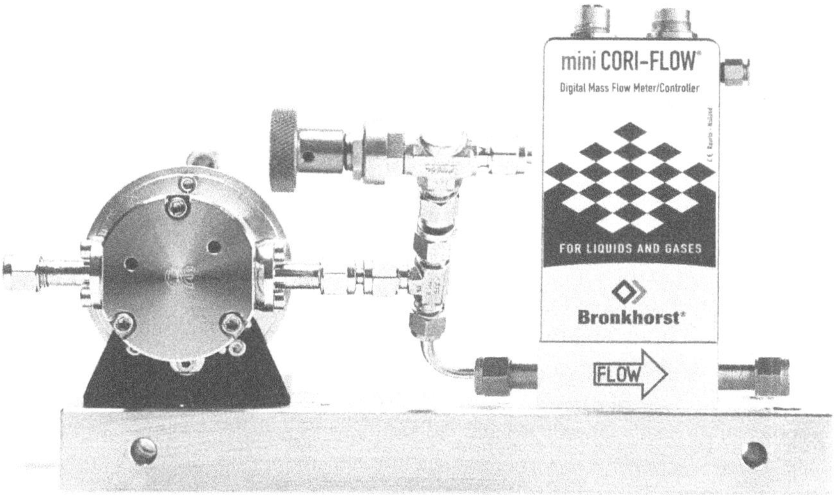

FIGURE 9.3 Bronkhorst's mini CORI-flow MFC.

The company's CORI-FLOW precision mass flowmeters are based on the Coriolis principle and incorporate a single sensor tube. They offer better accuracy and are faster than thermal mass meters and can be used for both liquids and gases. All CORI-FLOW models include an IP65 weatherproof housing and are optionally available with ATEX approval for use in Zone 2 hazardous areas. They also can be equipped for bidirectional flow measurement. CORI-FLOW mass flowmetering instruments provide high accuracy in two classes: ±0.2% of reading for liquid and ±0.5% of reading for gas. Bronkhorst also offers a line of portable dosing systems using a compact design that includes a Coriolis or ultrasonic flowmeter, a gear pump, an operation module, and a power supply.

FRONTIERS OF RESEARCH

The Use of Mass Flow Controllers Will Expand in Industrial Segments, Including Those Involving Alternative Energy, Solar Power, and Energy Storage

Strong growth areas in industrial markets for MFCs include the adoption of hydrogen as a clean energy source. Uses of hydrogen include controlling hydrogen gas in fuel cells and electrolyzers. MFCs are used in solar/photovoltaic cell manufacturing processes. MFCs are also essential to battery production, where they are used for controlling gas flows in battery production and energy storage technologies.

The industrial markets are made up of a wide variety of industrial segments. Industrial markets such as automotive are going through a transitional period as uncertainty in consumer tastes becomes more of a challenge to predict. Others, such as emissions monitoring and the life sciences, are thriving due to environmental regulations and upgrades in manufacturing processes. The need to find alternative energy sources will also drive additional research and development in the foreseeable future. MFC manufacturers will benefit from these industrial segments as MFCs are relied upon to measure and control gas flows in these segments. Fuel cells, as one example, represent a more recent growth area for MFCs. This includes the broader category of alternative energy.

Keeping up with the Increasingly Complex Demands from the Semiconductor Industry

The ongoing push for more powerful, and more sophisticated semiconductors necessitates highly accurate and reliable MFCs for various processes. These include liquid precursor delivery to vaporizers, and ensuring the accurate delivery of gases like etching, carrier, and protective gases, which are critical for producing flawless microchips. Another critical process is chemical vapor deposition. The semiconductor industry is continuing to expand, with ever more complex and sophisticated processes. A major frontier of research for the mass flow controller market is keeping pace with these increasingly complex processes.

Miniaturization and Low Cost Present Challenges to MFC Manufacturers

The semiconductor and industrial markets are demanding even smaller and lower-cost MFCs. The low cost aspect is market-driven and depends in part on finding lower cost components. The trend towards miniaturization in the semiconductor industry is driven in part by the need to allow for a greater number of transistors to be placed on a given area of a silicon wafer, leading to increased processing power and functionality. This leads to faster processing speeds and reduced power usage. Miniaturization enables the creation of smaller, lighter, and more portable electronic devices, like smartphones, laptops, and wearables.

Diagnostics and Predictive Maintenance Enhance the Life of MFCs

MFCs benefit from predictive maintenance enabled by advanced diagnostics. These diagnostics monitor the MFC's health in real-time, predicting potential issues before they cause downtime or process errors. By analyzing sensor data, valve performance, and other parameters, predictive maintenance systems can identify problems like valve friction, air leaks, or calibration drift, allowing for timely maintenance and preventing costly failures. Smart MFCs continuously monitor key parameters like sensor output, valve position, temperature, and pressure. Diagnostic systems analyze data to identify deviations from normal operating parameters, signaling potential issues. Based on the identified issues, the system can trigger alerts or alarms, prompting operators to take corrective action before a failure occurs. This is a broader trend in the flowmeter industry, especially including ultrasonic and Coriolis flowmeters.

BIBLIOGRAPHY

Flow Research, Inc., *The World Market for Mass Flow Controllers*, 4th Edition (Wakefield, Massachusetts: Flow Research, Inc., 2025)

10 Multivariable Flowmeters

OVERVIEW

There are two main types of flowmeters that can measure several parameters besides volumetric flow to calculate mass flow. These variables are static pressure, temperature, and density. These two main types of flowmeters are as follows:

- Differential Pressure (DP) flowmeters
- Vortex flowmeters

There are two other flowmeter types used to measure mass flow by measuring additional parameters, and these were discussed in Chapter 4:

- Ultrasonic flowmeters
- Turbine flowmeters

DIRECT VS. INDIRECT MEASUREMENT OF MASS FLOW

So far in this book we have examined Coriolis and thermal flowmeters, along with mass flow controllers, as ways to measure mass flow. Of these, only Coriolis flowmeters, along with mass flow controllers using Coriolis technology, can be said to measure mass flow directly. Measuring mass flow directly means measuring it without relying on parameters other than mass flow to measure the flow. These parameters are typically temperature, pressure, and density. Multivariable flowmeters rely on these parameters to arrive at a mass flow measurement. Even thermal flowmeters do not measure flow directly, because they take the heat coefficient of the gas whose flow they are measuring into account in order to determine mass flow.

The flowmeters in this chapter rely on the following formula for measuring mass flow:

$$M = Q * \rho$$

Here:

M = Mass flow
Q = Volumetric flow

Mass flow is calculated by multiplying volumetric flow times density. This idea was explained in Chapter 1.

USE OF DENSITOMETERS FOR MEASURING MASS FLOW

The flowmeters discussed in this chapter are volumetric flowmeters, and these meters use volumetric flow as a means to determining mass flow. One way to obtain mass flow is to determine density is to use a densitometer, and then multiply volumetric

DOI: 10.1201/9781003457855-10

flow times density. Densitometers operate by correlating some physical property of a fluid with fluid density. Densitometers operate using a variety of principles:

- Vibration
- Buoyancy
- Nuclear (Gamma ray)
- Acoustic
- Buoyancy (Float type)
- Ultrasonic

Someone might choose such a solution if a Coriolis meter is too expensive, or if the line size is too large to accommodate a Coriolis meter. Industries that use this type of solution include petrochemical, chemical, food & beverage, life sciences, and pulp & paper, including slurry processing.

MEASURING TEMPERATURE, PRESSURE, AND MOLAR MASS

Another way to measure mass flow is to measure temperature and pressure, determine the molar mass of the gas, and then use these values to determine density. These values can be combined with volumetric flow to determine mass flow.

MULTIVARIABLE FLOWMETER MEASUREMENTS AND MULTIVARIABLE FLOWMETERS

Measuring mass flow is significantly more involved than simply measuring volumetric flow. This is because the mass of a fluid varies with temperature and pressure. It is also dependent on the particular gas being measured, and the molar mass of that gas. To calculate mass flow, the flowmeter or flow computer needs to use as part of its input the molar mass of the gas whose flow is being measured. This means that to adequately determine the mass flow of a gas, the type of gas (e.g., air, oxygen, nitrogen, argon, etc.) has to be input into the flowmeter or flow computer.

One way to accomplish a multivariable measurement is to use a temperature sensor or transmitter along with a pressure sensor or transmitter to measure the temperature and pressure of the fluid whose flow is being measured. The outputs of these temperature and pressure devices are then output to a flow computer. The volumetric flow value is also sent to the flow computer. A flow computer can also store other important values that may be needed to calculate flow. Some of these may be input by the operator. They include the inner diameter of the pipe, the identity of the gas or gas mixture being measured, molar mass values for different gases, the compressibility of the gas (z), humidity, and elevation. If a flow computer is used to determine gas flow, it is making a multivariable measurement. However, this does not make it a multivariable transmitter. If these values are put inside a flow transmitter, and the transmitter makes the flow calculation, then the transmitter is called a multivariable transmitter.

DIFFERENTIAL PRESSURE (DP) FLOWMETERS

A DP flowmeter consists of a DP transmitter integrated with a primary element and has the capability of calculating flowrate based on the difference in pressure. DP flowmeters rely on a constriction placed in the flow line that creates reduced pressure in the line after the constriction. A primary element is used to create the constriction in the flowstream. A DP flowmeter also requires a means to detect the difference in upstream vs. downstream pressure in the flow line. While this can be done with a manometer, today's DP flowmeters use DP transmitters that sense the difference in pressure, and then use this value to compute flowrate.

When most flowmeters are sold, the transmitter and sensor are sold together. This is true for ultrasonic, vortex, Coriolis, turbine, and other types of flowmeters. All these flowmeters operate based on a correlation between flowrate, or mass flow, and some physical phenomenon. For ultrasonic flowmeters, it's the difference in transit time of sound waves sent across the pipe. For turbine flowmeters, it's the speed of the rotor. DP flowmeters also correlate flow with a physical phenomenon; the difference in pressure upstream and downstream from a constriction in the flowstream.

Where DP flowmeters differ from other flowmeter types is that the transmitter is still often sold separately from the primary element that creates the constriction in the flowstream, sometimes from different suppliers. Because a DP flow transmitter cannot make a flow measurement without a primary element, customers who purchase a DP flow transmitter without the primary element are not actually buying a DP "flowmeter." They don't have a DP flowmeter until they connect the primary element to the DP flow transmitter. **Thus, a DP flowmeter is considered to be a DP flow transmitter that is connected to a primary element for the purpose of making a flow measurement. If the DP transmitter is a multivariable transmitter, then the result is a multivariable flowmeter.**

In the past, the pressure transmitter companies sold DP transmitters and users ordered their primary elements separately. Now, however, a number of companies are selling DP transmitters already integrated with a primary element, such as an Annubar or an orifice plate. It is tempting to consider these the only true DP flowmeters, however, a better description is that these are DP flowmeters with an integrated primary element. If a customer assembles a DP flowmeter by connecting up a DP flow transmitter to an orifice plate or a Venturi tube from another source, the result is just as much a DP flowmeter as an integrated product.

THEORY OF DIFFERENTIAL PRESSURE MEASUREMENT

The theory behind DP flowmeters is that energy is conserved when flow passes across or through a constriction in the pipe. A more exact statement of this theory is known as Bernoulli's principle, which states that the sum of the fluid's static energy, kinetic energy, and potential energy is conserved across a constriction in the pipe.

Bernoulli's principle for incompressible fluids is as follows:

$$P + \frac{1}{2}\rho v^2 + \rho g h = \text{constant}$$

Where:

- P = Static **Pressure** (Pa)
- ρ = **Density** of the fluid (kg/m^3)
- v = **Velocity** of the fluid (m/s)
- g = **Acceleration due to gravity** (9.81 m/s^2)
- h = **Height** above a reference point (m)

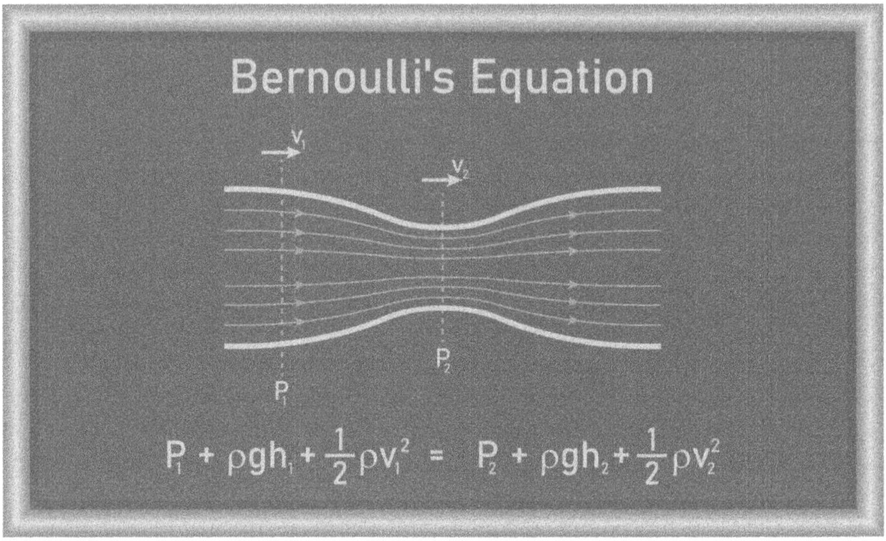

FIGURE 10.1 Bernoulli's equation.

Figure 10.1 displays this information in graphic form.

The equation of continuity formulates a relation between fluid flowrate and velocity for fluids that are incompressible. It can be formulated as follows:

$$Q = A_1 x\, v_1 = A_2 x\, v_2$$

Figure 10.2 An illustration of the continuity equation.

Q is volumetric flowrate. The product of A_1, the amount of an incompressible fluid that crosses the area at point 1 over some set amount of time, times v_1, the velocity at point 1, equals the product of A_2, the amount of fluid that crosses the area at point

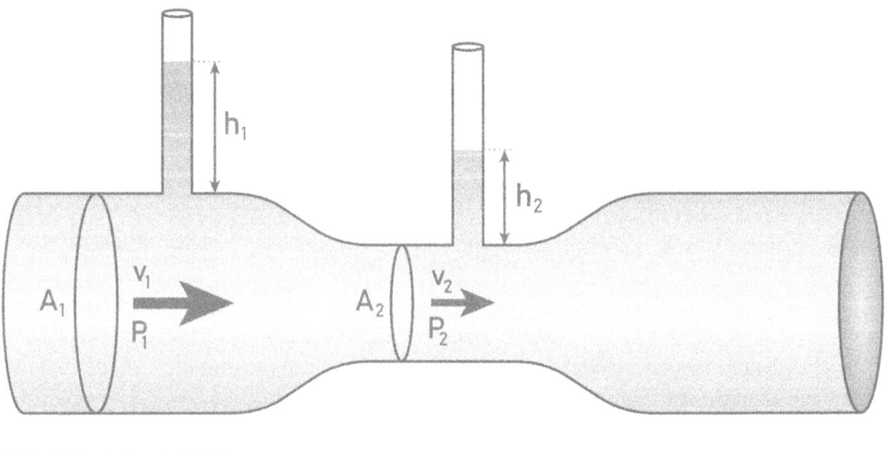

FIGURE 10.2 A DP flow measurement.

2 over the same set amount of time, times v_2, the velocity at point 2. (This is also sometimes written as $A_1v_1 = A_2v_2$.)

Combining Bernoulli's principle with the equation of continuity yields the result that the DP generated by an orifice plate is proportional to the square of the flow through the orifice plate. This is the "square root" relationship that is fundamental to all orifice plate and other DP flow measurements:

$$Q = k\sqrt{\Delta p}$$

In the above equation:

Q = volumetric flow, with units like m^3/s, L/min., etc.
k is a constant that is often determined with flowmeter calibration, but includes other factors such as the discharge coefficient and pipe geometry
Δp is the difference between upstream and downstream pressure, as measured by the transmitter based on the constriction in the flowstream created by the primary element

This formula contains the insight that volumetric flow is proportional to the square root of pressure drop. This holds for most primary elements, with adjustments for some.

TYPES OF PRIMARY ELEMENTS

The following types of primary elements are described below:

- Orifice Measuring Points
- Pitot Tubes
- Venturis
- Cone Elements
- Flow Nozzles
- Wedge Elements
- Combinations Elements (e.g., Nozzle/Pitot Tubes)
- Other (e.g., Dall Tubes, Laminar Flow Elements)

Orifice Measuring Points

Orifice plates are the most common type of primary element. An orifice plate is a flat, usually round piece of metal, often steel, with an opening or "orifice" in it. The orifice plate needs to be positioned at a correct position in the flowstream for it to function as a primary element for the purpose of making a DP flow measurement. To be in position, it must be held in place. This is typically done by an orifice assembly, an orifice flange, or a holding element.

In addition to an orifice plate and assembly or flange, most orifice plate installations require the presence of a valve manifold, which serves to isolate the pressure transmitter from the process. DP flow transmitters use either a three-valve or a five-valve manifold.

Since an orifice plate cannot serve as a functioning primary element unless it is held in proper position, and since valve manifolds are required for most DP flowmeter measurements, this chapter defines an orifice measuring point as having the following three components:

- Orifice plate
- Orifice assembly, flange, or holding element
- Valve manifold

Pitot Tubes

The Pitot tube is named after Henri Pitot, who invented it in 1732. Henry Philibert Gaspard Darcy, another Frenchman, published a paper in 1858 that made improvements in Pitot's invention. The first patent for the use of a Pitot tube to measure velocity in pipes was given to Henry Fladd of St. Louis, Missouri, in 1889. Pitot tubes are of two types:

- Single port
- Multiport averaging Pitot tubes

A **single-port Pitot tube** includes an L-shaped tube measuring impact pressure. This tube is inserted into the flowstream, with the opening facing directly into the flow. Another tube measuring static pressure has an opening parallel to the direction of flow. Flowrate is proportional to the difference between impact pressure and static pressure.

A **multiport averaging Pitot tube** has multiple ports to measure impact pressure and static pressure at different points. The DP transmitter computes flowrate by taking the average of the differences in pressure readings at different points.

Some companies such as Emerson Rosemount and Veris (now Armstrong International) have introduced proprietary versions of the averaging Pitot tube. Emerson Rosemount's proprietary version is called the Annubar, and it was formerly sold by Dieterich Standard, now part of Emerson Process Management. Veris's averaging Pitot tube is called the Verabar.

Venturi Tubes

The Venturi tube was invented by an Italian physicist named Giovanni Battista Venturi in 1797. In 1887, Clemens Herschel used Venturi's work to develop the first commercial flowmeter based on it. His version of the Venturi flowmeter became known as the Herschel Standard Venturi. Herschel published his paper called "The Venturi Water Meter" in 1898. In 1970, a company called BIF introduced the Universal Venturi Tube™.

A **Venturi tube** is a flow tube that has a tapered inlet and a diverging exit. The DP transmitter measures pressure drop and uses this value to calculate flowrate.

CONE ELEMENTS

Cone meters have been around since McCrometer, Inc. developed and patented the first successful model, its V-Cone®, in 1985.

Cone meters consist of a specially tapered element positioned within the flowstream to create a restriction to the flow through the pipe. Some cones are held in place by a tube that also connects them to the outside of the pipe. In the even more compact Wafer Cone® design, the element is fastened in place with an attached bar that spans the pipe diameter. The cone element creates a difference in speed and pressure as the flow is forced around it, then allowed to resume unobstructed flow beyond the cone. DP transmitters get measurements via a port in the pipe upstream from the cone (the high side) and also downstream (the low side) via another port either in the pipe wall or in the blunt end of the cone. Flowrate calculations are based on the same physical principles as other DP-type flowmeters, including Venturi meters. However, cone elements and meters are distinct from Venturi ones in design and are called just "cone" elements and meters. Perhaps the confusion arose from the "V-" in the name people became familiar with from the first and still successful cone meters, McCrometer's proprietary V-Cone? But McCrometer says that "V-" actually refers to the shape.

Cone meters can be used to measure gas, steam, and liquids in a wide variety of industries. Extremely robust, accurate, economical, low-maintenance, and useful in tight fits, they can be found from subsea to processing plants to Navy ships to satellites.

FLOW NOZZLES

A flow nozzle is a flow tube with a smooth entry and a sharp exit. With, it might be said, elegant simplicity, the restriction caused by the reductions in diameter within the nozzle creates measurable changes in the flow through it, but the smooth, tapered shape causes less permanent pressure loss than an orifice plate.

The DP transmitter computes flowrate based on the difference between upstream pressure and downstream pressure.

Flow nozzles are mainly used for high-velocity, erosive, non-viscous flows. Flow nozzles are sometimes used as an alternative to orifice plates when erosion or cavitation would damage an orifice plate. They offer excellent long-term accuracy.

WEDGE ELEMENTS

A wedge element is a flow tube that has a "V-" shaped flow restriction – the "wedge" – protruding into the flowstream from at least one side of the pipe. The wedge might be attached and contained inside the pipe, or it might be created by heavy sections of metal welded into a notch cut in the pipe. In the latter case, there are often reinforcing bars across the notch (parallel to the length of the pipe). In any case, the wedge is solidly built in, making these elements extremely robust, with no moving parts, no critical areas to wear or shift. They can handle any type of flow profile, whether laminar, transitional, or turbulent, and they are less prone to clogging or buildup than some other element types. The wedge shape is usually symmetrical – presenting the same on the upstream and downstream sides – thus allowing for bi-directional flow. Or, in the case of highly erosive fluids, should there eventually be significant enough wear on one side of the wedge, the flow tube can be turned around to present the unworn side to the flow, thus prolonging the life of the meter. They are easy to install and easy to use.

There are variations of wedge elements designed to handle air and gases, steam, and all sorts of liquids – clean, dirty, high-solids, slurries, viscous, corrosive or erosive.

GROWTH IN THE USE OF MULTIVARIABLE DP FLOWMETERS

Multivariable DP flowmeters measure more than one process variable, usually DP and/or pressure and temperature. Many of them incorporate a pressure and a temperature transmitter into a single device, making it unnecessary to order these products separately. Multivariable DP transmitters are mainly used to measure mass flow, and thus, they can be used for steam and gas flow measurement.

Multivariable DP transmitters require a primary element to create the restriction in the flowstream that enables the DP flow measurement. They can be used with orifice plates, averaging Pitot tubes, Venturi tubes, and other primary elements. They are increasingly popular as end-users seek to reduce costs, and also to do more flow measurement of steam and gas. Another attractive feature of these instruments is that they also provide more dynamic information about the process, enhancing management control and improving quality outcomes.

PARTNERSHIPS ARE EXPANDING THE REACH OF PRIMARY ELEMENTS COMPANIES

Partnerships and acquisitions are having a major impact on the flowmeter market. Examples include Emerson's purchase of Flexim, the partnership between Endress+Hauser and SICK AG. On May 13, 2025, ABB announced an agreement to assemble and sell RMG's gas flowmeters in the United States. One reason that companies are partnering up is the applications are becoming more complex, and it is becoming more difficult for a single company to meet so many different application needs.

In June 2023 TMCO, a supplier of orifice fittings, announced it would begin assembling and selling FMG's ultrasonic flowmeters in the United States. FMG is based in the Netherlands. It was founded by Ge Fischer, who also founded Instromet. This is an interesting example of a partnership because, as a supplier of orifice fittings, TMCO is steeped in conventional flowmeters. The partnership with FMG gives them an opportunity to compete in the new-technology flowmeter market. TMCO also sells FMG's turbine flowmeters. According to recent discussions with TMCO, this partnership has worked well for them. This is another example of cross-pollination in the flowmeter market, one that involves a primary elements company. For an interview with Ge Fischer, go to www.worldflow.com and look for Legends of Flow.

FRONTIERS OF RESEARCH

The following are frontiers of research for DP multivariable flowmeters.

Integrating Primary Elements with DP Transmitters to Create a Multivariable Flowmeter

The performance of DP flowmeters has been increased with the introduction of a primary element integrated with a DP transmitter. In Emerson's Annubar Averaging Pitot Tube flowmeter product, a primary element is integrated with a DP flow transmitter to create a DP flowmeter. This reduces the need for impulse piping and valves, and also makes it possible to calibrate the device before shipping with the primary element already attached. Emerson also produces the ProPlate, which is a DP transmitter integrated with an orifice plate.

The popularity of these integrated flowmeters is likely to increase as end-users seek to cut costs and simplify the installation process. Since these integrated flowmeters require a primary element, this is a growth factor for primary elements.

Improving Primary Elements Technology

Primary elements suppliers continue to develop technological improvements, many of which are created as solutions to rather discrete application problems. For example, Primary Flow Signal stocks and supplies 35 different Venturi elements, many of which have been developed for a specific level of performance, addressing customer applications that were not previously considered applicable for DP measurement.

Dosch Messapparate now includes averaging Pitot tubes in its primary elements portfolio as well as a new Cone meter and wedge elements that conform to ISO 5167-5 and ISO 5167-6 standards, respectively. Canalta Controls offers new Single and Dual Chamber Housings that ease the task of flow conditioner inspections and make these inspections more cost effective. Another recent addition to their portfolio is an internal valve isolation repair and/or installation kit. And this plug and play solution is suitable for use with all Canalta and Daniel chamber fittings, representing a sizable market base.

Expanding the Types of Orifice Plates

Emerson and other companies have brought out new types of orifice plates in the past five years, and this remains an area for further research. Types of orifice plates include concentric, eccentric, segmental, and quadrant edge orifice plates. These orifice plates have different shapes and have their openings located in different places. Conditioning orifice plates reduce the need for upstream piping. In some cases, these are offered with an integrated pressure transmitter, to form an integrated DP flowmeter.

Materials of construction are an important part of the effectiveness of orifice plates. Materials like nickel and Monel make the plates more corrosion-resistant. The different designs of the orifice plates can enhance accuracy and can also make the plate less susceptible to clogging by suspended particles in the flowstream. Even though orifice plates are probably the most studied type of primary element, new materials and new designs are a frontier of research for these very traditional forms of measurement.

Combining Several Technologies to Create a New Type of Primary Element

While most types of primary elements fit into a single category, some companies have innovated by combining two types of primary elements into a new type of primary element. This is somewhat like a compound meter, which combines a positive displacement and a turbine meter into a single type of flowmeter. Compound meters are used in situations where the meter has to measure high flows sometimes and low flows at other times. For example, they are used in apartment buildings where water flow is high in the mornings and evenings, but very low in the middle of the night.

Armstrong Veris offers the Accelabar, which is a combination of a flow nozzle and an averaging pitot tube. VorTek has introduced the VorCone, which combines elements of a vortex flowmeter, with elements of a cone meter. The tests and results on the VorCone are very positive. This combination of technologies makes it possible to predict the fluid density, volumetric flowrate, and mass flowrate without making use of any fluid density information from an external source. Among other benefits, this combination of technologies makes it possible for the VorCone to calculate the changing densities of gas mixtures.

While these are two examples, this is hardly the end of the story. Combining two technologies together can help compensate for the weaknesses of a technology and enable it to address a broader range of applications. For example, vortex meters have a difficult time measuring low flows, since if flow is low enough, the bluff body in the vortex meter will not generate the vortices required for a flow measurement. Pairing a vortex meter with a thermal meter could result in a meter that could handle medium to high flows and also excel at low flow measurement. While this is one theoretical example, there are no doubt other possibilities that will continue to be researched by innovative flowmeter companies.

VORTEX MULTIVARIABLE FLOWMETERS

Overview

Vortex meters are among the most versatile types of flowmeter. They can readily measure liquid, gas, and steam flow. Other new-technology meters such as ultrasonic and Coriolis can readily measure liquids and gases, but have difficulty with steam, and thermal meters are used almost exclusively for gas flow measurement. Magnetic flowmeters excel at measuring liquids but cannot measure steam or gas flow. Vortex meters are the only new-technology meter that readily measures all three fluid types.

Vortex flowmeters are well-suited for measuring steam flow, and they are widely used for this purpose. Steam is the most difficult fluid to measure due to the high pressure and high temperature of steam, and because the measurement parameters vary with the type of steam. Main types of steam include wet steam, saturated steam, and superheated steam. Steam is often measured in process plants, and is used for power generation. In addition to their ability to tolerate high process temperatures and pressures, vortex meters have wide rangeability. This allows them to measure steam flow at varying velocities. In process and power plants, steam is often measured coming from a boiler.

History

Vortex flowmeters were first introduced by Yokogawain 1969. Yokogawa's meter was designed to measure flare gas. Since that time, other major suppliers have come into the market, including Endress+Hauser and Emerson Rosemount. Other companies with a significant market presence in vortex include KROHNE, ABB, Foxboro, and VorTek Instruments.

Like many other flowmeters, vortex meters experienced application issues early in their development. One problem for vortex meters was the effects of vibration, which tended to create false readings. Suppliers created software to deal with vibration issues, and this proved to be an effective solution.

How They Work

Vortex flowmeters operate on a principle called the von Karman effect. This principle concerns the behavior of fluids when an obstacle is placed in the path of flow. Under the right conditions, the presence of the obstacle generates a series of alternative vortices called the von Karman street. This phenomenon occurs in liquid, gas, and steam, and has been observed in many diverse contexts including cloud layers passing an island and whitewater rapids.

In vortex flowmeters, the obstacle takes the form of an object with a broad, flat front called a bluff body that is mounted at right angles to the flowstream. Flow velocity is proportional to the frequency of the vortices and the flowrate is calculated by multiplying the area of the pipe times the velocity of the flow.

In order to compute the flowrate, vortex flowmeters count the frequency of vortices generated by the bluff body using a variety of techniques. The majority of vortex flowmeters use a piezoelectric sensor to detect vortices, however some use a capacitive or ultrasonic sensor (Figure 10.3).

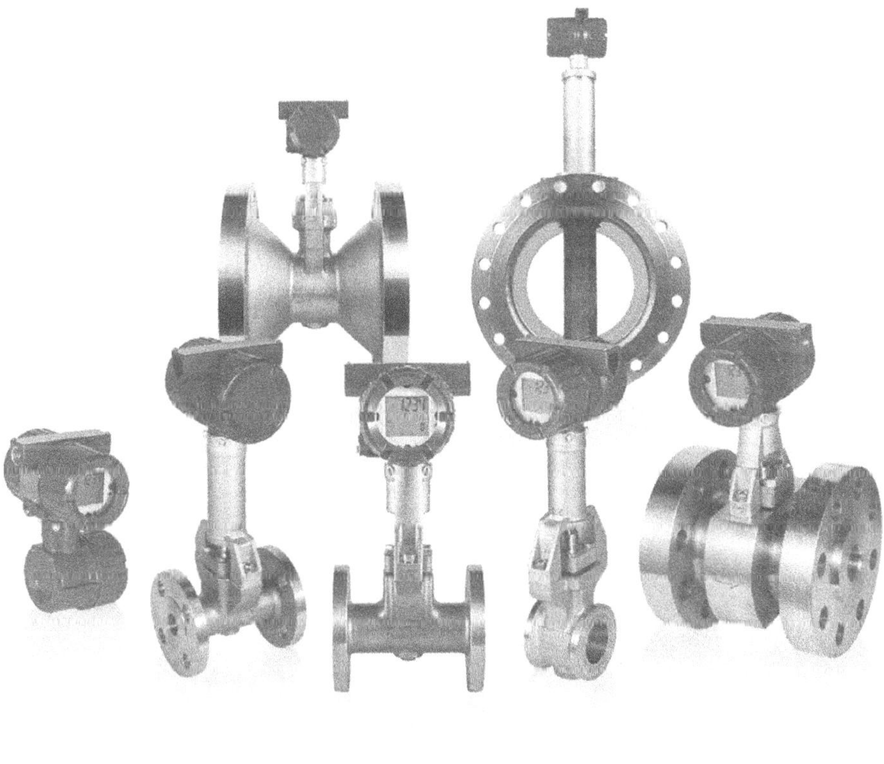

FIGURE 10.3 Illustration of Yokogawa's reduced bore type vortex meter, the digitalYEWFLO.

THE ORIGIN OF MULTIVARIABLE VORTEX FLOWMETERS

The origin of the multivariable vortex meter goes back to VorTek Instruments. Jim Storer had the idea of a multivariable vortex meter in 1994 while at EMCO Flow Systems (Figures 10.4 and 10.5). At that time, EMCO was selling vortex flowmeters along with pressure and temperature sensors and sending the values to a flow computer to calculate mass flow. Jim Storer had the idea to consolidate all these measurements into a single unit to create a multivariable vortex meter. In 1995, he left EMCO to found VorTek Instruments in Longmont, Colorado, where he began the design of the multivariable vortex flowmeter.

Storer lacked a distribution channel and for this purpose, he formed a joint venture with Sierra Instruments in Monterey, California. According to the terms of the agreement, Vortek retained the rights to and would manufacture the insertion multivariable vortex, while Sierra would own and manufacture the inline version. However, Sierra would be allowed to market both meters under its brand name. This arrangement lasted for several years until VorTek developed a second-generation multivariable vortex flowmeter, which Sierra sold under their brand name. This arrangement continued until about 2014, when Sierra decided to manufacture its own vortex meter. Even after this, Sierra continued to sell VorTek meters under its own name. This arrangement continues to this day. Now both companies are members of the TASI Group.

FIGURE 10.4 EMCO Flow Ssystems.

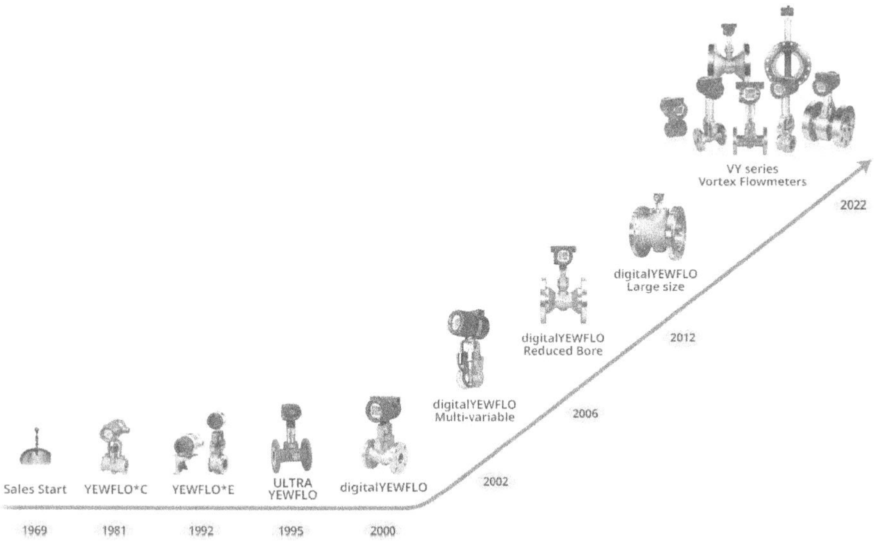

FIGURE 10.5 Timeline of Yokogawa vortex flowmeters.

In 2017, Jim Storer retired from VorTek and turned the reins of the company over to Eric Sanford, who served as president until early in 2025. Beginning in late 2012, VorTek Instruments was a subsidiary of Azbil North America, a wholly owned subsidiary of the Japan-based Azbil Group. On January 22, 2024, Azbil Corporation announced that it had transferred the Colorado-based Azbil VorTek to Sierra Instruments, now a company in the TASI Measurement portfolio (Figure 10.6).

Since multivariable vortex meters were first introduced, a number of other companies have come out with their own multivariable vortex flowmeters. These include ABB, Yokogawa, KROHNE, and Endress+Hauser. While multivariable flowmeters are somewhat more expensive than their single-variable counterparts, they enable users to obtain significantly more information about the process than a single-variable volumetric meter. This additional information can result in increased efficiencies that more than make up for the additional cost of the multivariable flowmeter. Multivariable vortex flowmeters also have the capability of measuring mass flow, and this makes them attractive, especially for steam and gas flow measurement.

Multivariable flowmeters measure more than one process variable. In addition to process fluid flow, multivariable vortex flowmeters have one of the following three sensor configurations:

- Integrated temperature sensor; no pressure sensor
- Integrated temperature and pressure sensors
- Integrated pressure sensor; no temperature sensor

About two-thirds of multivariable vortex flowmeters are shipped with an integrated temperature sensor but with no pressure sensor.

FIGURE 10.6 An M22 Inline Vortex Flowmeter from VorTek Instruments

IMPORTANT ADVANTAGES

Vortex meters have some disadvantages when compared to other new-technology flowmeters. They are not as accurate as Coriolis meters, they are more intrusive than ultrasonic meters, and are less widely used than magnetic meters. They are more expensive than DP flowmeters. Yet despite these comparisons, vortex flowmeters offer a number of important advantages when compared to both new-technology and traditional technology meters.

Vortex flowmeters offer accurate and reliable flow measurement at a competitive price. The average worldwide selling price of vortex meters can be less than a third of the average worldwide selling price of Coriolis meters and less than a quarter of the average worldwide selling price of ultrasonic flowmeters. Even though vortex meters are not as accurate as Coriolis meters, many vortex meters offer accuracy readings of better than 1%, depending on fluid and application. While vortex meters are somewhat intrusive compared to magnetic and ultrasonic meters, they are much less intrusive than the orifice plates used with DP transmitters to measure flow. Pressure drop from vortex meters is minimal, since most shedder bars are relatively small in size. While vortex meters are less widely used than magnetic flowmeters, this is due in part to the fact that they were introduced nearly 20 years after magnetic flowmeters.

Vortex flowmeters offer an accurate and reliable means to measure flow. They cannot achieve the high accuracy of Coriolis flowmeters, but neither can any other type of flow measurement device. The most common applications do not require the absolute highest accuracy available and can be suitably addressed with that offered by vortex flowmeters. Vortex measurement accuracy can be as high as ±0.65% for liquids and ±0.9% for gases and steam, with repeatability of ±1.0% of rate for liquids, gases, or steam.

Vortex meters can handle a wider range of process conditions than almost any other flowmeter. Vortex meters can readily handle liquids, gases, and steam. While Coriolis meters are well known for their ability to handle liquids, they are less widely used on gases, and their use on steam is very recent. Ultrasonic meters were used in the 1980s to measure gas flow but are only beginning to be used for steam. By contrast, the use of vortex meters to measure steam flow is well established. Vortex meters can handle the high pressures and temperatures associated with steam flow measurement.

FRONTIERS OF RESEARCH

Vortex technology will continue to play a critical role in industrial markets. Applications will continue to demand reliability, dependability, and low-maintenance flow solutions, and digital protocols will continue to evolve in industrial instrumentation. These trends present an opportunity for innovation in vortex Ethernet solutions – including easy-to-use Ethernet-based protocols – and connectivity capabilities. This wider bandwidth will enable vortex transmitters to deliver more diagnostics and process insights. In addition, as measuring value information about plant processes and status grows increasingly important, customers are requiring smart vortex meters that supply much more information for predictive and condition-based maintenance, self-diagnostics, big data, Internet of Things, and more.

Other frontiers of research for vortex flowmeters include:

- Measuring Low Flowrates
- Dual-Sensor Flowmeters
- Vortex Flowmeter Performance

MEASURING LOW FLOWRATES

One weakness of vortex technology is the difficulty that vortex flowmeters have in measuring low flowrates. This difficulty is inherent in the technology itself. Shedder bars may not shed vortices at all, or they may not shed vortices at a regular rate if the flow is too low. Other meters, such as thermal mass meters, do a better job than vortex meters of measuring flow at low flowrates.

In a no-flow situation, a vortex meter will register zero flow. However, it is unable to immediately respond as flow begins. A vortex meter with an output of 4–20 mA will have an output of 4 mA at zero flow. As flow picks up, it may suddenly jump to 4.8 mA when the flowrate is fast enough to register on the vortex meter. The flowrate that is represented by the readings between 4 mA and 4.8 mA is a "blind spot" for the meter.

Some companies have taken steps to improve the capability of their flowmeters at low flowrates by using a very small shedder bar that is more sensitive to lower flowrates than meters with larger shedder bars. While this technology can push back the limits of low flow, it does not completely solve the problem of low flow measurement.

Another approach to handling low flowrates is through the use of reduced bore meters. Reduced bore vortex flowmeters are narrower in diameter where the bluff body generates vortices than at either end of the flowmeter. Generally, the internal diameter of the meter is reduced by either one meter size or two meter sizes. Flow speed is increased where the pipe is narrowed. Reduced bore vortex flowmeters can more easily handle low flowrates. Reduced bore meters are also typically easier to install than standard bore meters, and their installation costs tend to be less.

While reduced bore vortex flowmeters have only appeared fairly recently, beginning with the single line size versions, they are becoming increasingly popular. Look for more suppliers to introduce them in the next several years, including both one-meter and two-meter size reduced bore meters. Their use will increase as more customers learn about their advantages, and have a chance to try them out on their own applications. Emerson and Yokogawa are among the companies that manufacture reduced-bore vortex meters.

DUAL SENSOR FLOWMETERS

Another innovation in vortex flowmeters is in dual sensor meters. When a shedder bar generates vortices, a sensor downstream detects the vortices and the flowmeter counts how many there are. These sensors are typically piezoresistive, ultrasonic, or capacitive. A recent innovation is to put two sensors downstream rather than just one, for the purposes of redundancy. This makes for a more reliable flowmeter, though it doesn't necessarily enhance performance.

A second innovation is to have two vortex flowmeters in the same line and calibrate them together. This is also done for the purpose of redundancy. It is reminiscent of the arrangement some companies have for custody transfer of natural gas where they put two turbine meters or two ultrasonic meters in series. While having two vortex flowmeters in series does not improve the performance of the individual meters, it does provide a more stable and reliable installation.

VORTEX FLOWMETER PERFORMANCE

The vortex flowmeter market has not grown as rapidly as the Coriolis and ultrasonic markets, and it is not nearly as large as the magnetic flowmeter market. One factor holding it back is that it is more intrusive than those other three meters. A bluff body can get knocked out of place, as can an orifice plate, and the result is an unreliable measurement. Coriolis, ultrasonic, and magnetic flowmeters are not intrusive in this way, and they provide more stable measurements.

Second, there is a limit to how much information the vortex transmitter can derive from counting the number of vortices. Apparently this is not sufficient for vortex flowmeters to achieve accuracies of much over 0.5%. Ultrasonic and Coriolis meters reach higher levels of accuracy. In order to achieve a breakthrough in accuracy or performance, someone will have to figure out how to derive additional information from the vortices than is currently being derived.

APPROVAL FOR USE IN CUSTODY TRANSFER APPLICATIONS

Custody transfer of natural gas is a fast-growing market, especially with the increased popularity of natural gas as an energy source. Natural gas changes hands, or ownership, at a number of points between the producer and the end-user. This changing of hands occurs at custody transfer points, and it is an application that is highly regulated by the American Gas Association (AGA), the American Petroleum Institute (API), and other approval authorities internationally. In January 2007, an API committee approved a draft standard for the use of vortex flowmeters for custody transfer of liquid and gas. This approval has provided some boost in sales of vortex meters over time, as suppliers develop products that conform to the standard. The standard was first updated in 2010 and then reconsidered in 2013. It was finally adopted in 2017 as a standard by the API for custody transfer of gas flow measurement using vortex flowmeters.

Even though it is taking some time for the API approval to have a noticeable effect on the vortex flowmeter market, this approval provides manufacturers with another avenue of approach to potential vortex flowmeter customers. It is likely this some company or companies will come out with "custody transfer" versions of this flowmeter, meaning vortex meters specifically designed in accordance with the API specification.

BIBLIOGRAPHY

Flow Research, Inc., *The World Market for Vortex Flowmeters*, 8th Edition (Wakefield, Massachusetts: Flow Research, Inc., 2025)

Flow Research, Inc., *The World Market for Primary Elements*, 3rd Edition (Wakefield, Massachusetts: Flow Research, Inc., 2025)

David W. Spitzer, *Industrial Flow Measurement* (Research Triangle Park, North Carolina: Instrument Society of America, 1990)

11 The Rope Experiment and Finite Point Geometry

OVERVIEW

This ground-breaking chapter takes a radical look at mathematical concepts inherent in flow measurement and proposes alternative theories that avoid the contradictions of traditional geometry. We explore geometry from a new perspective and develop the fundamentals of a new geometry called Finite Point Geometry. In this new geometry we show how to derive a rational value for calculating the circumference and area of a circle rather than treating pi (π) as an irrational number that has an infinitely non-repeating decimal. In the process we challenge the father of geometry, Euclid, and his foundational belief that points have location but no extension or area, and we provide a solution to the resulting Zeno's Paradox. In addition, through a new spin on calculus, we offer a fresh approach to tackling one of the most difficult tasks for a traditional geometry based primarily on straight lines: measuring the area under a curve.

We dive deeply into these concepts in the following sections:

1. An overview of geometry as a branch of mathematics
2. Euclid's views on points and the number line
3. Euclid's views on points and lines generate Zeno's Paradox and other paradoxes
4. The prevailing solution to Zeno's Paradox
5. Where Zeno goes wrong
6. Measurement requires a unit of measurement and a level of precision
7. Points lie on the line, not in the line
8. How many points lie on the line
9. From length to area: Measuring the area of a circle
10. Why we need π in the first place
11. The role of π in calculating the area of a circle
12. The Rope Experiment
13. The length of a rope doesn't change when you lay it flat
14. The area of a circle is a rational number
15. Measuring flow in round pipes
16. Application to mass flow
17. Calculus and measuring the area under a curve

We begin by placing geometry in the context of mathematics, then discuss the obvious but important fact that making a measurement requires a unit of measurement

DOI: 10.1201/9781003457855-11

and a level of precision. We explore the relationship between points and lines, and propose that points lie *on* the line, not *in* the line.

We look back on how π originated and the various attempts to determine its value. A diagram shows that r^2 is a square and that the formula for the area of a circle, $\pi * r^2$, is essentially finding how many squares will fit into a circular area. We conclude that the answer appears to be no finite number.

The Rope Experiment section then shows that the value of the circumference of a circle is rational. The experiment essentially takes the circumference of a circle and lays it into a flat line, which of course doesn't change the length of the rope. As a flat line, the length can be measured with a ruler or tape measure, and there is no need for π. We can measure the value of π as 3.1416, or we can carry out the value to any level of precision. The value 3.1416 is not the beginning of an infinite series of nonrepeating numbers; it is a rational number that is determined by measurement.

If we substitute C/d for π in the formula for the area of a circle (πr^2), we get $(C/d) * r^2$. Because C is now rational by the rope experiment, and we know that the diameter is rational in most situations, it follows that the area of a circle is a rational number. This is a dramatic change from the traditional interpretation of πr^2, according to which the area of a circle is an irrational number with an infinitely non-repeating decimal.

GEOMETRY AS A BRANCH OF MATHEMATICS

In many ways, mathematics is the language of science. Mathematics displays abstract relationships that may not be apparent to an ordinary observer. None of the tools and objects we rely on for our daily life, such as planes, trains, automobiles, computers, clocks, flowmeters, and cellphones would be possible without mathematics. Mathematicians, scientists, and philosophers have spent many centuries developing different branches of mathematics including arithmetic, geometry, algebra, calculus, trigonometry, number theory, probability and statistics. Political polling, bridge building, housing construction, traffic control, space travel, and many other features of our daily lives rely on different branches of mathematics.

Geometry is one of the most central and fascinating branches of mathematics. Geometry is the branch of mathematics that studies the sizes, shapes, properties, and relationships of points, lines, surfaces, solids, and fluids. Geometry studies not only the dimensions of the objects of our ordinary experiences, it also describes relationships among abstract entities that these objects represent. For example, there are many square and rectangular objects in the world. Geometry can tell us how to find the areas of those objects by using formulas that apply to any object of that shape. For example, to find the volume of a rectangle, multiply:

$$\text{Length} \times \text{Width} \times \text{Height}.$$

This is a formula that applies to any rectangle, regardless of what it's made of, its size, color, height, etc. Geometry gives us abstract formulas and equations that apply to many different objects of a certain shape, independent of their other qualities. This is part of the power of geometry.

Much of our geometry goes back to the early Greeks, in the period around 300 B.C. This is the time when Euclid lived. Euclid is sometimes called "the father of geometry." He earned this title through his rigorous axiomatic method, and his geometric proofs. Euclid is the author of "Elements," a 13-book work that lays out his definitions, axioms, theorems, and proofs, along with his study of the properties of lines, triangles, circles, squares, and other geometric shapes. It is to him that we owe our concept of point. Euclid's definition of "point."

Euclid's definition of "point" is "that which has no part." This means that a point has no length, width, or breadth. A point has location only and does not have size or dimension. This is the concept of a point that Euclid laid out around 300 BC, and it is still the dominant definition of a point 2,800 years later.

EUCLID'S VIEWS ON POINTS AND THE NUMBER LINE

The distance between two points is often represented as a line. This line can be straight or curved, depending on the location of the points. Many attempts have been made to define the continuum that is a line. The traditional Euclidean view is that a continuous line is made up of discrete points, and that another point can be placed between any two points. This means that a line is dense, and there is no limit to the number of points that can be drawn in the line. In fact, in the Euclidean view, a line is made up of infinitely many points. These points are conceived of as having location only, with no area.

The relation between discrete points and continuous lines typifies some mathematical puzzles that have been discussed for thousands of years and are still discussed today. Continuous lines and discrete points are fundamentally different in nature and it is not clear whether it can be correct to analyze continuous lines as being made up of discrete points, even if there are very many of them. Here the idea of infinity is brought in to be a kind of metaphysical glue. The idea is that if there are infinitely many discrete points, then these points, when packed densely enough, form a continuum, which then somehow adds up to continuity. But it is not clear how arealess points can somehow make up a continuous line, even if there are infinitely many of them.

A similar problem occurs with analyzing the area under a curve. This problem is addressed by calculus, which was independently invented by Newton and Leibniz in the late 1600s. According to calculus, this area is determined by drawing a series of rectangles that more and more closely approximate the curved area. As these rectangles become smaller, the sum of their areas more closely approximates the curved area. As the widths of the rectangles, which become more and more numerous, approach zero, the sum of the areas of the rectangles equals the area under the curve. This sum is conceived of as the limit of an infinite series of rectangles that are ever diminishing in width.

EUCLID'S VIEW ON POINTS AND LINES GENERATES ZENO'S PARADOX AND OTHER PARADOXES

Analyzing a continuous line as being made up of infinitely many discrete points isn't necessarily wrong, and this idea has been accepted for thousands of years. It does raise some paradoxes, however, which need to be dealt with in some way. Some

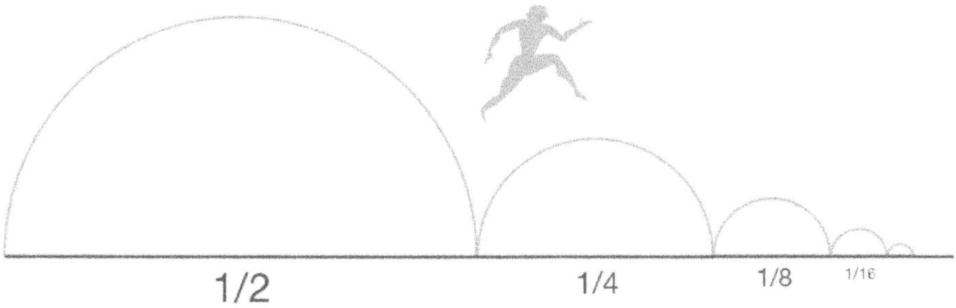

FIGURE 11.1 An illustration of Zeno's paradox.

of these are called "paradoxes of motion," and they were formulated by a Greek philosopher named Zeno of Elea around 450 B.C. It is believed that Zeno preceded Socrates by about 20 years.

Zeno may have formulated as many as 40 paradoxes, but only 10 are known. One involves Achilles and a tortoise, and it gives an analysis of motion that purports to show that Achilles, a fast runner, can never catch up to the tortoise, provided the tortoise is given a head start.

Another paradox, well enough known to be called Zeno's Paradox, has a similar logic and involves the idea that motion is impossible. To understand this paradox, consider someone attempting to move from Point A to Point B. First, he must go halfway from A to B, which we can call A′. To move from A′ to B, he then has to go halfway again to B, to a point we can call A″. However, he is still not at B, since he must go halfway again, ad infinitum. So, no one ever actually gets to point B, and hence motion is impossible (Figure 11.1).

Consider the concept of a point as a dimensionless mathematical object with no area, which is based on the Euclidean view of the number line. As discussed previously, in Euclidean geometry, a line is made up of infinitely many arealess points. It is always possible to put another point between any two points.

What makes Zeno's Paradox seem plausible is that the person who is going from Point A to B is conceived of as being located at an arealess point. Since it is always possible to put another point between any two points, he never reaches destination B.

THE PREVAILING SOLUTION TO ZENO'S PARADOX

What is the correct solution to Zeno's Paradox? The prevailing contemporary solution relies on concepts from set theory developed by Bernard Bolzano, Richard Dedekind, Georg Cantor, and others. According to this view, Zeno's series of points is no longer considered to be potentially infinite. Instead, it accepts the idea of an infinite set of points and analyzes Zeno's series of points as an infinite set of points that converge to a limit. Instead of viewing the set of path lengths or points that is 1/2 + 1/4 + 1/8 + 1/16 + etc., as an infinity that can never be traversed, this analysis views it as a series that converges to 1. This idea of a set of points that converge to a limit is similar to the idea of a series of rectangles of diminishing width that converge to a limit of zero and is used in contemporary calculus to analyze the area under a curve.

WHERE ZENO GOES WRONG

Where does Zeno's paradox go wrong? This paradox only works because the person involved is considered to be at a dimensionless point. Once the person is considered to be at a mathematically dimensionless point and at a position not taking up area, it is not possible to refute Zeno's paradox. This is because it will always be possible to specify a dimensionless point between the point where our traveler is located and his endpoint.

INFINITY, A KIND OF METAPHYSICAL GLUE

Euclidean geometry neglects a simple mathematical distinction in analyzing what a line is. According to this geometry, points do not lie *on* the line, they lie *in* the line. A line, as we've said, is conceived of as being made up of infinitely many dimensionless points. Conceiving of points as lying in a line requires the introduction of the concept of infinity, since there appear to be infinitely many points in a line. Set theorists have also created different infinities of different sizes. The concept of infinity is a kind of metaphysical glue that is required to analyze a continuous phenomenon such as a line in terms of discrete arealess points. There is no way to build up to the concept of continuity when it is being analyzed in terms of discrete points without introducing the concept of infinity.

MEASUREMENT REQUIRES A UNIT OF MEASUREMENT AND A LEVEL OF PRECISION

When people make a length measurement, they are implicitly relying on a unit of length measurement, whether it is inches, miles, meters or kilometers. And that measurement unit has to be defined in terms of a standard for that unit, otherwise the measurement unit cannot serve as an objective standard of measurement. When the human foot was the standard by which the length of a foot was judged, but there was no objective foot length to serve as the standard, the foot unit referred to different lengths. When a "standard" foot length was inscribed at St. Paul's Church in the 1100s, this was a first step toward defining an objective standard for the length of a foot.

The concepts of length and distance are closely related. Length is the measurement of the distance between two points. As long as the measurement is being made in feet or inches, then feet or inches are the smallest allowable units of measurement for the purpose of this measurement.

Every measurement is made to some degree of precision. The term "precision" refers to the number of digits to which a measurement is made. A measurement is not absolute; it is relative to the unit of measurement and to the degree of precision required for the measurement.

While there is no theoretical limit to the degree of precision of any measurement, this does not mean that the degree of precision attainable is infinite. It is simply unlimited in the sense that it can be taken as far as the need requires and technology allows. It is not infinite in the sense that the precision of measurement can go to infinity. One way that Zeno's Paradox goes wrong is that it involves constantly shifting the level of the precision regarding where a person is located.

POINTS LIE *ON* THE LINE, NOT *IN* THE LINE

Anyone who pays attention to our language will realize that we speak of points being on a line more naturally than of points being in a line, or in part of a line. The idea that points are in a line is more a result of mathematical analysis than of understanding the language of mathematics. But what is the difference between points being on a line and points being in a line?

Someone who is sitting on a fence is not part of the fence; instead, his body is physically touching the fence. But no one would think that a person sitting on a fence is part of the fence. Instead, the fence is made up of wood, steel, rocks or some other material, depending on the type of fence. Likewise, a book lying on a table is not part of the table, although the book touches the table.

Intuitively, it makes sense to say that points lie on a line. When we draw a point on a line, the line is typically there first, and we physically mark the point on top of the line. We might mark the point with an "X" (X marks the spot), with a round dot or with a little perpendicular line. However, it is marked, it would be unusual to conceive of this "X," round dot or perpendicular line as somehow being a part of the line, while it is perfectly natural to think of the X, round dot or perpendicular line as lying on the line.

If points lie on the line rather than in the line, there is no need to introduce the concept of infinity to describe how many points there are. This is because a group of points lying on a line are not continuous; instead, they are simply a group of discrete points that are related by all being on the same line. It is the line that is continuous, but since discrete points are not part of a continuous line, they in themselves are not part of the continuous phenomenon.

A LINE IS THE PATH OF A MOVING POINT

What is the relationship between points and a line? As Aristotle says in De Anima 1:4, a line is the path of a moving point. Likewise, a plane is the path of a moving line. A point and a line are then intimately related, but not in the way Euclidean geometry thinks of them as being related. A line is somewhat like the trail of a meteor, except that when we draw a line using the point of a pencil, the line is static and remains visible.

HOW MANY POINTS LIE ON A LINE?

If we cannot analyze a continuous line as an infinite set of arealess points, how should it be analyzed? We have already said that points lie on the line but should not be considered part of the line. In this analysis, it makes sense to think of points as discrete units that sit on the line. If it is the number line that is being considered rather than a line that consists of the distance between two points, then these points can be considered as dimensionless in the Euclidean fashion. Euclidean mathematicians will consider these as representing an infinite number of points. In a non-Euclidean analysis, these points will have area. This area will vary with the unit of measurement that is specified. The Euclidean analysis doesn't reintroduce Zeno's Paradox with points lying on the line, as long as these are not conceived as representing the location of three-dimensional objects.

If we conceive of points as having area, no matter how small, and as sitting on the number line rather than being a part of it, then there is no possibility of introducing Zeno's Paradox. For example, consider how many points exist between 0 and 1 on the number line. If each point is defined as having a width of 1/1000 of an inch, where the distance between 0 and 1 is considered to be one inch, then there are 1000 points between 0 and 1. If someone is considered to be located at one of these points, Point A (for example, point 200), and wants to go halfway to Point B, which is 1, he will go to point 600. From there, to again go halfway to Point B, he will go to point 800, then to point 900, then to point 950, and on to Point 975. In Zeno's Paradox, he would next go to point 987 ½. However, this move is not allowed since there are no half points in this system. Zeno's Paradox is prevented if points are considered as having area or width, no matter how small. Once their area or width is defined, then it is not possible to place another point between any two points on this line.

Pursuing this non-Euclidean conception of a point as having area or width, no matter how small, it is always possible to redefine the width or area to any level of precision desired. For example, a point can be defined as one/millionth of an inch, or one-trillionth of an inch. Then there are a million points between 0 and 1, or a trillion points. These points are conceived as touching each other at the edges so there is no room for additional points once their size is determined. And since fractions of a point are not allowed, Zeno's Paradox cannot be generated. If a smaller or larger point is required, then the unit of measurement needs to be defined accordingly.

FROM LENGTH TO AREA: MEASURING THE AREA OF A CIRCLE

Most people are familiar with the formula for measuring the area of a circle: $\Pi * r^2$

Here 'r' stands for the radius of the circle. While this formula may look simple, unpacking what it means has vexed thousands of people, including many great minds, over the centuries. This section takes a look at the problem of defining π.

WHY WE NEED Π IN THE FIRST PLACE

The history of π goes back more than 4,500 years. Around the year 2000 B.C., the Babylonians and the Egyptians had attempted to determine the value of π. Since these early times, there have been many efforts to calculate π. Since that time, the value of π has attracted the minds of many great thinkers over the centuries, including Gauss, Newton, and the incredibly prolific Swiss mathematician Leonard Euler. Around 1779 Johann Lambert proved that π is irrational. In 1882, F. Lindemann proved that π is transcendental, thereby proving the impossibility of squaring the circle.

Clearly it is important to know the value of π, and trying to understand and determine this value has attracted the attention of some of the greatest minds over the past 4,500 years. But how did the need to know π come about in the first place? Knowing the value of π is the key to knowing the area of a circle. This area is given by the formula $A = p * r^2$.

The early Babylonians and the Egyptians found it impossible to measure the length of the circumference of a circle as a round area. Instead, they tried approximating its length with polygons and other straight line shapes that closely resembled the distance around the circle. This approach remained the prevalent one for many centuries. Eventually, with the invention of calculus and infinite series, the focus turned to finding a mathematical equivalent to the circumference of the circle. These efforts proved fruitless as well.

Once the computer age arrived in the 1960s, the focus of those seeking a value for π turned to calculating it to as many decimals as possible, in hopes of finding a set of repeating digits. On Π Day, March 14, 2024, a California storage company in California called Solidigm announced that it had calculated π to 105 trillion digits. The calculations took 75 days to complete and used up 1 million gigabytes of data. If you type this number out in 10-point type on a piece of paper, it would be 2.3 billion miles long. This broke the previous record set by Google Cloud in 2022, which had calculated π to 100 trillion digits.

THE ROLE OF π IN CALCULATING THE AREA OF A CIRCLE

Figure 11.2 shows why there is a need to use π in calculating the area of a circle:

The area of the square with four sides each equaling r is obviously less than the area of the circle whose radius is r. More significantly, the formula requires that some number of squares fit into the circle with radius r. Yet just as you can't fit a square peg into a round hole, seemingly no definite number of squares will fit into that circular area. The only number that works in this formula is π, an irrational nonrepeating decimal, according to traditional mathematics.

The diagram above shows the difficulty in using a square as a unit of measure for the circle. It was believed that the value of π as an irrational number can never be determined because there is no rational number of squares that will fill the area of a round circle. Also, it shows the difficulty in arriving at an exact value for π, since it is seen as the ratio of the circle to the diameter, and people couldn't find a way to measure the length of the circle exactly.

Because π was defined as a ratio of Circumference/Diameter, people tried to find a ratio that equaled π. Examples include 22/7 and 355/113. Leibniz and others even created

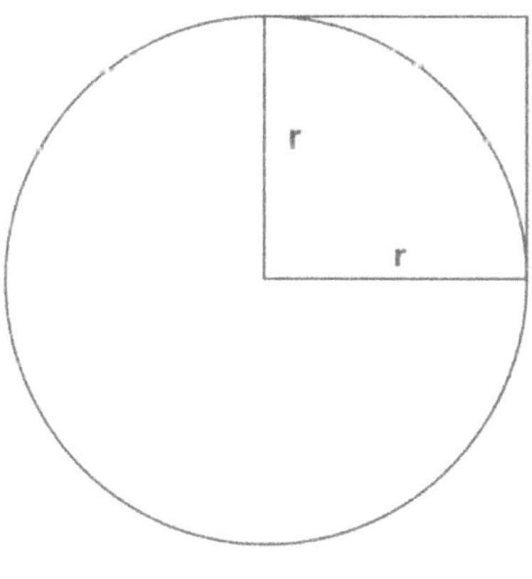

FIGURE 11.2 The value r^2 as a unit for measuring the area of a circle.

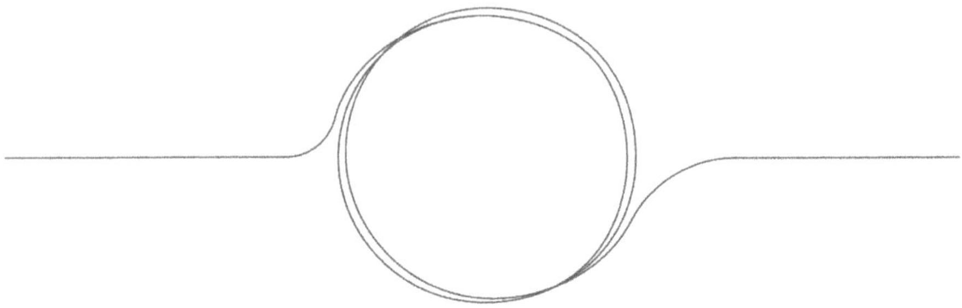

FIGURE 11.3 A rope being formed into a straight line.

convergent series that closely approximated π. The problem is that none of these ratios or convergent series equaled π exactly. So it was assumed to be an irrational nonrepeating decimal number. Calculating π to 105 trillion digits is an admirable feat of technology, but it gets us no closer to understanding π than does the number 3.1416.

THE ROPE EXPERIMENT

The Rope Experiment provides a different way to measure the length of the circumference of a circle (Figure 11.3).

Take a circle with a radius (r) of 2 inches and a diameter (d) of 4 inches. By our standard equation for the circumference (C) of a circle, the circumference is: $C = 2 * \pi * r$

Given that our radius is 2 inches long, this gives the following value for the circumference of the circle: $C = 2 * \pi * 2$

Multiplying this out, we have: $C = 4 * \pi$

Given that π is approximately 3.1416, this gives us a value of 12.5664 inches for the circumference of the circle.

Now take a rope and form it into a circle with a radius of 2 inches. This can easily be done by measuring in a straight line from the center of the circle to its edge, which is formed by the rope (Figure 11.4).

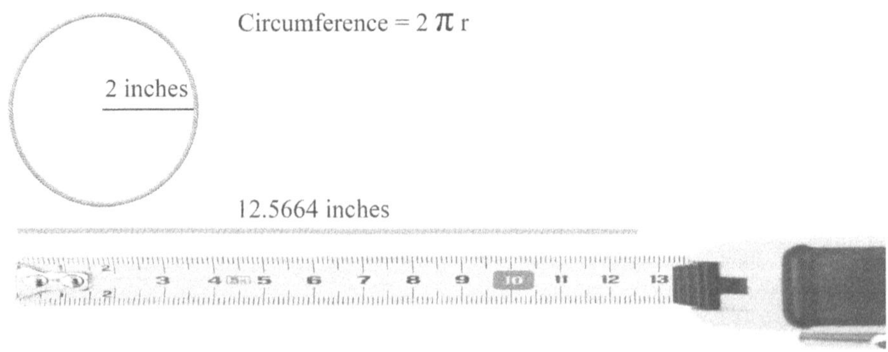

FIGURE 11.4 Measuring the length of a circumference of a circle with a tape measure.

ILLUSTRATION BY LESLIE BUCHANAN

Take the rope that forms the circle and lay it flat. It becomes a straight line of length C. A straight line has measurable, rational length (you can measure it with a ruler or counting units). There is no need to use π in providing the length of a straight line. So C, the circumference, is a rational number.

Measure the rope with a yardstick or a tape measure. You will find that the length of the rope is right around 12.5 inches, depending on how precise you want to be. Using the approximate value of π as 3.1614 in the formula 2 * π * r, the circumference is 12.5664 inches. What is the difference between 12.5 inches and 12.5664 inches?

Using the traditional value of π for the circumference in equation 2 * p * r, we get an irrational value that can only be approximated because it includes a value (π) with indefinitely many digits that never repeat. Π has been carried out to 105 trillion digits by supercomputers and it never repeats digits.

Using a tape measure, we get a measurable value for length of the rope that is just over 12.5 inches. We can carry this out to any desired level of precision, but it depends on our unit of measurement and the precision of our measuring tool. Given that we are using inches as our unit of measurement, it is unlikely that we would need to carry it out much further than four digits after the decimal. Whether we use a fraction or a decimal value, the measured number we get is a rational number and it does not contain a never-ending string of decimal values.

Which value is correct? Between 12.5 and 12.5664, we can say that 12.5664 is a more precise version of 12.5: the same measurement, but one carried out to a greater number of digits.

THE LENGTH OF THE ROPE DOESN'T CHANGE WHEN YOU LAY IT FLAT

A rope doesn't change its length when you shape it into a circle, turn it into a hula hoop, shake it up and down, or tie it around a tree. The length of the rope remains constant regardless of its shape. The ratio of a circle's circumference to its diameter is defined by the length of the circular boundary measured in rational units. The result is a rational value, specific to the precision of measurement, not a transcendental abstraction.

Given that we have a measurable, rational length for the rope when it is laid flat, this is the length of the rope. And this is also the length of the rope when it is shaped into a circle. The problem is not with our measurement of the rope – the problem is with our equation for calculating the circumference of a circle.

As a result, I suggest we accept that the value for the circumference of a circle is a rational value. When we measure the length of the rope as 12.5664 in a circle with radius 2, and the length of the circumference is 12.5664, this gives us the equation 12.5664 = 2 * π * 2. This equation is equivalent to 12.5664 = 4 * π. Dividing both sides by 4, we have 3.1416 = π. Since we know that the circumference of a circle is a rational, measured value, it follows that π is also rational. Since 3.1416 is the closest rational approximation to the value of pi, we can accept 3.1416 as the, rational value of pi obtained by measurement.

Using our formula:

$$C = 2 * \pi * r$$

we get

$$\pi = C/(2 * r) = C/D$$

where D is the diameter of the circle.

In our example: $\Pi = 12.56/(4) = 3.14$

This is close to the standard value for π of 3.1416

If we carry this division out to two more digits of precision: 12.5664/4 = 3.1416

This is the value of π carried out to the first four digits after the decimal point.

Rather than trying to find a single value of π that applies to all situations, we may want to approximate π as closely as necessary given the precision we require in our measurements. If we are carrying our measurement to only two decimal places in our measurement like in the example a value of 3.14 may do just fine. If we need more precision we can measure to a greater number of decimal places. Whatever we do, π remains a rational number, and C/D gives a rational value for π. Some people use 22/7 as a value for π. While this is less precise than 3.1416, it may serve well in some contexts.

The reasoning using ratios and converging series was based on defining π indirectly – through limits and approximations – rather than measuring it directly. The Rope Experiment offers a different approach: by measuring the length of a circular path laid flat, we find that π is a rational, measurable number. Now that we know from The Rope Experiment that π is a rational number, it is time to reexamine our assumptions about how to calculate the area of a circle.

THE AREA OF A CIRCLE IS A RATIONAL NUMBER

The formula for the area of a circle is $\pi * r^2$. If we substitute C/D for π in this equation, we get $(C/D) * r^2$. Because C is now rational by The Rope Experiment, and we know that the diameter is rational in most situations, it follows in this framework that the area of a circle is a rational number. This is a dramatic change from the traditional interpretation of $\pi * r^2$, according to which the area of a circle is an irrational number with an infinitely non-repeating decimal.

Having a rational value for the area of a circle means that we can reconcile the unit square with circular area. Earlier in this chapter, we stated that measurements require a unit of measurement and need to be carried out to some degree of precision. We can apply this idea to the area of a circle: since we have so many digits to choose from, we can carry the calculation of area to as many digits as we like, by measuring the circumference to as many digits as we like, taking C/D, and multiplying this value by the radius squared (r^2).

The value of 3.1416 … as an irrational number was derived by multiple methods. One method was to inscribe and also circumscribe multiple polygons around a circle until the number of polygons approximates the circle, and then divide this value by the diameter of the circle. Another way is to multiply out 22/7. There are other methods that involve convergent series and limits.

In Finite Geometry, π is defined as the measured ratio of a circle's circumference to its diameter, with a rational value such as 3.1416. This measurement can be carried out to any finite degree of precision, but it is not shorthand for an irrational number with an unending series of digits. It is a real, physical quantity, like the meter, and it is based on measurable properties of space.

This method of measuring the circumference of a circle can be compared to the way in which the standard meter in Paris was used to define the length of a meter. Since 1983, this length has been defined by light speed instead of by the standard meter bar. But the idea is the same. We are advocating a physical method for determining the length of the circumference of a circle that generally results in a value of 3.1416. This value may differ according to the precision of the measurement tool. It can be carried out to any degree of precision desired, but it is not shorthand for an infinite series or for an irrational number. The length of the circumference of a circle is now a measurable constant derived from real geometry – the circumference of a circle with known diameter.

THE LENGTH OF THE METER IS ALSO DEFINED BY A PHYSICAL OBJECT OR PROCEDURE

Since 1855, the science of metrology has made major advances, with a greatly expanded capability for precision. In 1866, an Act of Congress made the use of the metric system permissible in the United States. In 1889, the United States received a prototype Meter bar from the International Bureau of Weights and Measures, and that became the new standard for the length of a meter. However, in 1960, the science of measurement took a major step forward with the definition of the meter in terms of the wavelength of light from krypton-86. A meter was defined as 1,650,763.73 wavelengths of the orange-red radiation of krypton-86. This definition was promulgated by the French organization called General Conference on Weights and Measures (CGPM).

In 1959, just prior to this definition of a meter, Australia, Canada, New Zealand, South Africa, the United Kingdom and the United States agreed on the length of an international yard, defined in terms of the meter. The international yard is equal to 0.9144 of a meter. Thus, the length of the yard came to be derived from the definition of a meter.

In 1983, the CGPM took another step toward precision by defining the meter as equal to the path length traveled by light in a vacuum during a time interval of 1/299,792,458 of a second. This definition also locks in the speed of light in a vacuum as 1/299,792,458 meters per second. History indicates, however, that this is probably not the final word on the subject.

MEASURING FLOW IN ROUND PIPES

The value of π also plays a crucial role in flow measurement. There is no escaping the need to measure the area of a pipe when calculating flow. The formula for measuring volumetric flow is:

$$Q = A * v$$

Here Q is the volume of flow that passes a specific point in a unit of time. A is the cross-sectional area of the inside of the pipe and v is the average flow velocity. Circular area is calculated using the formula:

$$\text{Area} = \pi * r^2.$$

Because pipes are round, they form a circular area. The formula for the area of a circle is

$$\text{Area} = \pi * r^2$$

This means that the formula for Area is a part of most calculations for the flowrate for fluid flowing through a round pipe. Engineers typically use 3.13, 3.1416, or 22/7 as the value for π. They are taught to believe that this gives them an approximation of the area, since we cannot truly know the value of π. But the result of The Rope Experiment is that we can know the actual value for area. The precision required dictates how many digits should be used in a given situation.

Since $\pi * r^2$ is the formula for the cross-sectional area of a pipe, it is important to understand what it represents. The Greek philosopher Plato taught us that mathematics is not about actual squares and circles, and other geometric shapes, but about the ideal shapes that these figures represent. Thus, for example, if the surface of a desk does not have perfectly straight edges, we still feel comfortable using the formula length × width to calculate its area. Another way of saying this is that geometry is about ideal shapes and figures even though the actual objects we are talking about are not ideally round or ideally square.

In the case of pipes that contain fluid to be measured, there are other issues that affect the measurement besides the fact that the mathematics assumes an ideally round pipe. If all pipes were perfectly round, then there would be no issue with using the formula for the area of a circle to calculate their area. But there are multiple factors affecting the roundness of a pipe.

When considering the area of a pipe, it is also important to take into account the fact that pipes are three-dimensional objects. Pipes have an outside diameter (OD) and an inside diameter (ID). It is only the inside diameter of the pipe when cross-sectional area is being calculated.

WHAT IS THE INSIDE DIAMETER OF YOUR PIPE?

Not all pipes are perfectly round. When flowmeters are calibrated, they are calibrated on clean pipes with no sediment or buildup on the walls. Once flowmeters reach the field, they enter a new world where sediment and buildup can occur. This can occur in many ways.

Scaling. When hard water is heated, it can scale when heated, causing a hard, chalky deposit in pipes. This is due to the high levels of calcium and magnesium in hard water. This scaling hinders water flow, reduces appliance efficiency, and increases energy costs. Over time, it narrows the pipe's diameter, increasing the chances of blockages and leading to more frequent maintenance and potential replacements. Chlorine and disinfectants in municipal water supplies can react with pipe materials, degrading the pipe's inner surface.

Corrosion. Corrosion in water pipes is the result of the interaction of water, oxygen, and other elements of a pipe's environment. Corrosion can weaken the interior and exterior of the pipes. Aging pipes are subject to corrosion when exposed to oxygen and water. This is especially true of pipes made from metals like iron, steel, or copper. This process degrades the material, forming rust and releasing sediments into the water supply. The weakening integrity increases the risk of leaks and can lead to significant water damage and costly repairs. Rust and other compounds can build up on the inner pipe wall, reducing the internal diameter of the pipe. This can restrict the flow and increase pressure loss.

Temperature Changes. Extreme temperature changes can cause minerals to precipitate and solidify, forming deposits on the inner surfaces of pipes. In hot water systems, heat can induce calcium carbonate precipitation. These deposits can reduce pipe diameter, increase pressure, and decrease water flow.

Inside diameter is especially important for ultrasonic flowmeters, which use the distance from one side of the pipe to the other in their computation of flowrate. While this can be an issue for inline ultrasonic meters, it is also important for clamp-on ultrasonic meters. Clamp-on ultrasonic meters base their flowrate calculations on a signal sent through the pipe wall and to the other side of the pipe. Buildup in the pipe can contribute to the attenuation of the signal. It can also affect the calculation of the flowrate.

Of course, there are many other issues providing uncertainty in flow measurement. These include flow profile, cavitation, turbulence, repeatability, two-phase flow, and many more. All these issues need to be addressed, and this is what manufacturers do in their research and development efforts, using the math that is available to them. Having a rational value for the area of a circle may reduce or eliminate one of those uncertainties.

APPLICATION TO MASS FLOW

In Chapter 4, we learned that we can derive mass flow from volumetric flow through the following equations:

$$\dot{m} = \rho \times Q$$

Here:

$\dot{m}$ = mass flowrate (kg/s)
ρ = fluid density
Q = volumetric flowrate (m^3/s)

To get the density measurement, you need:

- An external density measurement, such as with a densitometer, or
- A calculated density based on pressure, temperature, and fluid composition (especially for gases).

This is true for flowmeters such as differential pressure and vortex that provide mass flow by measuring volumetric flow and then using density values to determine mass flow. Coriolis and thermal flowmeters measure flow differently, even though they may also output volumetric flow. However, even Coriolis and thermal flowmeters need to know the pipe size, and hence the internal diameter of the pipe, in order to determine mass flowrate.

Since the cross-sectional area of a pipe is a critical part of most flowrate measurements, whether volumetric or mass, it follows that knowing a pipe's cross-sectional area is an important part of most volumetric and mass flow measurements. Having a rational value for calculating this area is another step towards producing accurate and repeatable flow measurements.

CALCULUS AND MEASURING THE AREA UNDER A CURVE

Besides measuring the area of a circle, measuring the area under a curve has proved one of the most difficult tasks for a geometry based primarily on straight lines. Calculus provides one way to measure this area. Calculus finds the area under a curve using the definite integral:

$$\int_a^b f(x)\,dx$$

The definite integral represents the limit of a sum of infinitely many rectangles, each with:

Height: f(x)
Width: dx (an infinitesimal)

Calculus draws a rectangle under the curve, where one point of the rectangle intersects with the curve. It then draws a series of increasingly small rectangles with width (dx). The area under the curve is the limit of the sum of this process of drawing ever-narrower rectangles. Calculus takes the limit of the number of rectangles as that number approaches infinity and as their width approaches zero.

While calculus does not claim to actually complete an infinite process, it does rely on the concept of infinity in other ways. In particular, calculus relies on the idea that the number line is made up of infinitely many arealess points. Otherwise the idea of "approaching a limit" makes no sense. Calculus relies on the idea that between any two points are infinitely many other arealess points. If calculus reaches a gap in the number line, its approach to the limit of a series stops. In calculus, the series of rectangles must come arbitrarily close to the limit, which means for any position the rectangles reach in relation to the limit, it must be possible for them to get even closer.

We saw above that the assumption that there are infinitely many points in the line leads to Zeno's Paradox and other paradoxes. How then should we evaluate the area under a curve?

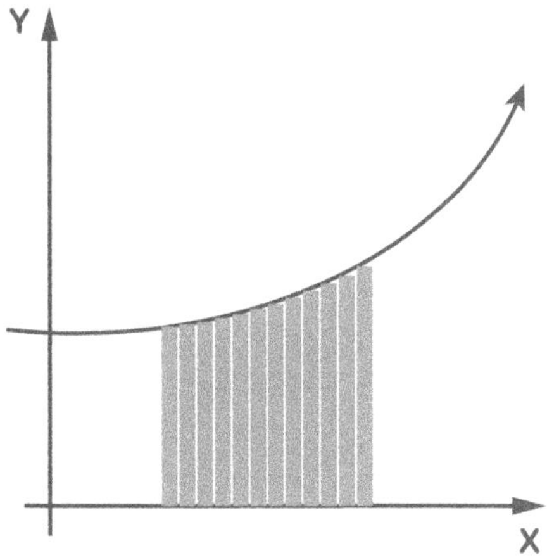

FIGURE 11.5 Drawing wide lines from the base to corresponding points on the curve to obtain the area under the curve.

FINITE POINT GEOMETRY AND THE AREA UNDER A CURVE

In Finite Point Geometry:

- Points lie on the line, not in the line.
- There is no limit to the number of points on a line, but each must have a specified area.
- Each point contributes a finite amount of area and serves as a unit of measurement.

Finite Point Geometry evaluates the area under a curve in the following way:

Consider a curve that extends two inches in width, say from 0 to 2.

Specify a number of points on a line that extends underneath the curve (e.g., on the x-axis). This implies a width for each point – in this case, 1/10 of an inch.

Draw a vertical line from each point along the line to a corresponding point on the curve.

Measure the lengths of all these lines and add them together. Divide that number by the number of points (in this case, 20). That is the average height of the curve. Now multiply the length of the line (2 inches) by the average height of the curve. This gives you the area under the curve.

This method does not require assuming an infinite number of points in or on the line. It is a method for constructing the area under a curve, and it provides a rational alternative to calculus (Figure 11.5).

BIBLIOGRAPHY

Petr Beckmann, *A History of π (PI)* (New York: St. Martin's Press, 1974)
Euclid, *The Thirteen Books of The Elements*, Translated with commentary by Sir Thomas L. Heath (New York: Dover Publications, Inc., 1956)

Appendix: The "Waterwheel" Myth and Coriolis' Actual Writings

THE COMMON CLAIM

Many modern summaries of Gaspard-Gustave de Coriolis' work state that his 1835 paper on rotating systems was inspired by waterwheels, or that he was talking about waterwheels in this 1835 article.

For example, Wikipedia (as of 2025) asserts:

Gaspard-Gustave de Coriolis published a paper in 1835 on the energy yield of machines with rotating parts, such as waterwheels.

This phrasing, repeated in numerous secondary accounts, is not supported by Coriolis's own writings. Although it provides an intuitive image of his work's mechanical background, it is not supported by the text in Coriolis's 1832 or 1835 papers.

WHAT CORIOLIS ACTUALLY WROTE

In his "Mémoire sur les équations du mouvement relatif des systèmes de corps" (Coriolis 1835), Coriolis' aim was theoretical. He writes:

> de déduire les équations générales du mouvement relatif des corps, et d'examiner la valeur du travail des forces dans ce mouvement.
>
> to derive the general equations of relative motion of bodies and to examine the work done by forces in this motion.

Throughout the memoir and in his earlier "Sur le principe des forces vives dans les mouvemens relatifs des machines" (Coriolis 1832), he refers broadly to systèmes de corps en mouvement relatif, machines en mouvement de rotation, and forces d'entraînement.

Nowhere does he use the word roue (wheel), roue hydraulique (waterwheel), or moulin (mill). His focus was on the geometry and mechanics of motion in a rotating frame of reference, not specifically about waterwheels.

ORIGINS OF THE "WATERWHEEL" INTERPRETATION

The link to waterwheels seems to have appeared in late-19th- and early-20th-century interpretations, when educators and engineers sought tangible examples of rotating machinery.

Industrial mechanics texts of the period often used hydraulic machines – particularly waterwheels – as examples to illustrate energy transfer in rotation.

Later commentators described Coriolis's work as "inspired by rotating devices such as waterwheels," almost certainly as a contextual inference rather than a quotation. Grenoble's INP history notes provide such an example.

Thus, the phrase "such as waterwheels" is an interpretive addition introduced by later authors, not a phrase found in Coriolis's text.

WHY THE DISTINCTION MATTERS

The difference between "rotating systems" and "waterwheels" is important. Coriolis's 1835 memoir was theoretical, extending Newtonian dynamics to non-inertial frames of reference.

The "waterwheel" retelling makes it sound empirical and device-specific, suggesting an engineer studying a mechanism rather than a mathematician formulating a general law. Recognizing this distinction restores Coriolis to his proper historical role: a theorist of motion, not merely an analyst of machines.

FROM WATERWHEELS TO MERRY-GO-ROUNDS

As physics education evolved, examples of rotating frames shifted from industrial machines to playground platforms. By the 20th century, the "merry-go-round" became the dominant analogy for the Coriolis effect, which was later generalized to the Earth's rotation.

The chain of analogies therefore, runs:

$$\text{rotating systems} \rightarrow \text{rotating machines} \rightarrow \text{waterwheels} \rightarrow \text{merry}-\text{go}-\text{rounds} \rightarrow \text{Earth}.$$

Each step made the analogy more understandable but distanced the concept from Coriolis's own intent.

Index

For Product Safety Concerns and Information please contact our EU
representative GPSR@taylorandfrancis.com
Taylor & Francis Verlag GmbH, Kaufingerstraße 24, 80331 München, Germany

www.ingramcontent.com/pod-product-compliance
Ingram Content Group UK Ltd.
Pitfield, Milton Keynes, MK11 3LW, UK
UKHW020205200926
475702UK00004B/629
9781032601601